AF522406

Encyclopaedia of
PLANT ECOLOGY

Volume 2

Shehzad Ahmad

ANMOL PUBLICATIONS PVT. LTD.
NEW DELHI - 110 002 (INDIA)

ANMOL PUBLICATIONS PVT. LTD.
H.O.: 4374/4B, Ansari Road, Darya Ganj,
New Delhi-110 002 (India)
Ph.: 23278000, 23261597
B.O.: No. 1015, Ist Main Road, BSK IIIrd Stage
IIIrd Phase, IIIrd Block,
Bangalore - 560 085 (India)
Visit us at: www.anmolpublications.com

Encyclopaedia of Plant Ecology

First Published, 2007

ISBN 978-81-261-3398-7 (Set)

PRINTED IN INDIA

Printed at Amit Enterprises

CONTENTS

Preface

Plant Ecology is a newer sub-discipline of Ecology. Plant Ecology is the study of plants and trees etc. in their environment, especially how they relate to each other. In fact, all the living beings including plants and their direct or indirect relation to the non-livings, constitute the ecosystem. For a healthy life, they all play vital roles and for a sound ecosystem, a few precautions, taken by us, are necessary.

Environment is the most significant factor for the existence of living beings, including plants. It includes climate, pollution, deforestation, conservation, disasters of flora and fauna and natural resources, etc. Not to say that, the present condition of the ecosystem is in a sorry state. With decreasing forests, rapid growth of industrialisation and globalisation etc., have left, in its wake, a number of other problems, challenging and threatening the whole existence of the vegetation and greenery on earth. All these issues demand our urgent heed. They have caused a huge loss to the natural balance. If we do not put a check on them, then different crises, looming large would become inevitable and incurable.

The prevailing condition makes it incumbent upon every one of us to try our level best to think over the ways; how to minimise the effect of these fatal factors. We should also be considerate enough to preach others, as well, to take the same initiative; components of ecosystem, *i.e.*, plants and animals could be saved from destruction for now and in future.

This work, *"Encyclopaedia of Plant Ecology"* is a precious gift to every one concerned with the subject. Prepared with utmost care, it would, undoubtedly, serve a great purpose.

All wise and positive comments from scholars, professionals and readers are welcome.

Editor

1

AWARENESS ABOUT ECOLOGY

- Leagans: *"Extension education is an applied science consisting of contents derived from researches, accumulated field experiences and relevant. Principles drawn from the behavioural sciences, synthesised with useful technology, in a body of philosophy, principles, contents and methods focussed on the problems of out of school education for adults and youths".*
- "Agricultural extension is a bridge that fills the gap between agricultural research stations on the one hand and the farming population on the other by the establishing a suitable teaching organisation at various levels of administration."
- "Extension education is a behavioural science following a continuous, persuasive and discriminating educational process, it aims at affecting the behavioural components of people in a desirable direction, through conviction, communication and diffusion by its Proven methods, principles and philosophies resulting in learning involvement of both client and change-agent systems."

PRINCIPLES OF COMMUNICATION

Leagans: *"It is a process by which two or more people exchange ideas, facts, feelings or impression in ways that each gain a common understanding of the message in essence it is the act of getting a sender and a receiver tuned together for a particular message or series of message"*

"Communication is the process of effecting an interchange of understanding between two or more people".

- Teaching and learning is the process of communication.
- Teacher acts as catalyst.
- Communication is a two way or double way process.
- Communication involves
  - (a) Source and his message
  - (b) Receiver and his response.
- The chief emphasis in communication on target audience.
- Communication systems

  K—Knowledge

  G—Generating

  S—System

  D—Disseminating

  C—Consuming

  Agencies

  Universities
  Research institutes
  Private organisation
  Voluntary organistion } KGS

- ***Mass Media***

  Universities
  Research Institutes
  Govt. agencies
  Voluntary organisations } KIDS

- ***Farmers***

  In agriculture KCS

Key Elements of Communication

(a) Sender

(b) Message

(c) Message treatment *i.e,* encoding

(d) Communication medium

(e) *Receiver:* It is the most important element because every element is directed towards receiver.

(f) Receiver's response.

Communication Models

Sender—Message—Treatment —Receiver—Channel

- Leagan's Model

Communicator—Message—Channel—Response of —Audience —Treatment — Audience

- According to Berlo

Source—Encoder—Message—Receiver—Decoder—Channel

Consequences of Communication

Communication media and extension methods are the tools of approach as they increase the effectiveness of extension work.

- An extension worker, must first know *"what methods are available to* him" secondly, he should know-"When *to* use a given *method* and thirdly *"he should become efficient in using each method"* The approaches are given below
- Approaches in communication and extension

(A) Individual approach through:

(1) Personal visits

(a) Farm visits (b) Home visits

(2) Personal letters

(B) Group approach through

(1) Demonstration

(a) Method (b) Result

(2) Training the leaders

(3) Discussion meetings

(a) Group discussion (b) Panel discussion

(c) Symposium (d) Forum

(e) Group interview (f) Workshop

(g) Dialogue or public conversation

(4) Field days

(5) Tours

(C) Mass approach through

(1) *Films and Slides:*

(a) Farm visits, (b) Home visits

(2) Flash-cards (3) Printed materials

(4) Radio (5) Recorded talks

(6) Charts, diagrams, etc. (7) Circular letters

(8) Newspaper stories (9) Compaigns

(10) Television

Division According to Equipments

(1) *Projected Aids:* Here projection is controlled by motor and electricity, *e.g.*,

(a) Film Projector

(b) Slide Projector

(c) Overhead Projector: It is indirect projection and transparencies used.

(d) *Opaque Projector:* There is no need of transparency.

(2) *Non-projected Aids:* Here no machinery is required, *e.g.* Flash cards, groups, charts, leaflets, folders, black boards.

Division According to Audio-visual Aids

(1) *Audio:* Radio, taperecorder

(2) *Visual:* Flash cards, overhead projector, slides, etc.

(3) *Audio-visuals:* Puppet, drama, television, cinema, etc.

MECHANISM OF COMMUNICATION

Individual

(1) Farm and Home Visit or personal contact

(2) Letters (Personal)

(3) Telephone call

(4) Office call

(5) Result demonstration

Group Contact Method: Group means a group of 2-30 persons:

(1) Symposium (2-5 speakers)
(2) Pannel (2-8 speakers)
(3) Discussion
(4) Lecture
(5) Tour and field days
(6) Method demonstration and result demonstration
(7) Meeting
(8) Role playing
(9) Puppet show
(10) Slide
(11) School
(12) Conference
(13) Seminar
(14) Training camp
(15) Public address equipment
(16) Flash card

Mass or Community Contact Method

(1) Bulletins
(2) Newspapers
(3) Journal
(4) Magazine
(5) Exhibition
(6) Circular letter
(7) Fairs
(8) Posters
(9) Cinema (Film)
(10) Movie
(11) Drama and songs
(12) Radio
(13) Leaflet/ folder
(14) Television

Glossary

- *Aim:* Aim is a broad objective
- *Audio-visual Aid:* Audio-visual aid is an instructional device that can be heard and seen.
- *Audience:* Audience is the intended receiver of messages.
- *Association:* It is a group of persons organising themselves for the purpose of fulfilling certain interests.
- *Adopter:* Adopter is a person who has continued full use of innovation.
- *Andragogy:* It means art and science of helping adults to learn.
- *Agricultural Extension:* Agricultural extension is the assistance to farmers to help them to identify and an their production problems and to become aware of the opportunities for improvement.
- *Audio-aids:* Audio-aids are those which facilitate understanding of a problem or topic through hearing.
- *Agricultural Information Centre:* Agricultural information centres are set up at district head quarter. It collects information suitable to the farmers.
- *Black Boards:* Black board is one of the most widely used visual aids which can be easily prepared everywhere with the use of locally available material and resources.
- *Calender of Work:* It is a plan of work arranged chronologically according to the time, when each step of work is to be done.
- *Circular Letter:* Circular letters are those which are written by persons who are at the high helm of affairs such as C.M., important public leaders.
- *Communication:* Communication is a process by which two or more people exchange their ideas, facts, feelings or impressions in ways that each gains a common understanding of the meaning and use of messages.

- *Channel:* Channel is the physical bridge between the sender and the receiver of a message.
- *Cooperatives:* These are the institutions organised by groups of individuals for mutual help, thrift and benefit.
- *Community:* Community is a sub group having many of the characteristics of society but on a smaller scale and with less intensive and co-ordinated common interest.
- *Continuous Education:* Continuous education can be given through both formal and informal types of education as and when need arises to educate for new technological development and for socio-cultural development.
- *Contract Farmer:* A farmer who act as intermediary between follower farmers and the extension education agency.
- *Costoms:* Costoms mean uniformly accepted ways of acting about some social aspects of life.
- *Caste:* When class is somewhat hereditary it is called as caste.
- *Culture:* Culture is the sum total of the ways in which human beings live and is transmitted from generation.
- *Charge Agent:* A professional person who attempts to influence adoption decisions in a direction he feels desirable.
- *Demonstration Centres:* These provide effective communication about improved methods of farm practices to the farmers by practical displays.
- *Decision Making:* Decision making refers to the process in which programmes are prepared by various departments in consultation with various persons.
- *District Agricultural Officer:* District agricultural officer is a person who makes research findings known to farmers for their use.

- *Dramatics:* Dramatics are a great source of entertainment and education and have been known to have a considerable effect on the society from times immemorial.
- *Distance Education:* It is any one of the various forms of the study which are not under the continuous immediate supervision of tutors present to their students in lecture rooms.
- *Demonstration:* Demonstration is conducted by farmer under direct supervision of extension workers to prove the advantages of recommended practices.
- *Demonstration Plot:* It is a small area set aside for result demonstration of extension staff.
- *Extension:* Extension is a method of non-formal education aiming at inducing behavioural changes in its clients for their increased knowledge, skills and income.
- *Education:* Education is the process of bringing desirable changes in human behaviour.
- *Extension Education:* Extension education is an applied science consisting of content derived from research, accumulated field experience and relevant principles drawn from behavioural sciences, synthesised with useful technology in a body of philosophy, principles, contents and methods focussed on the problem out of school education for adults.
- *Extension Worker:* Men and women employed by an extension agency or technical and professional staff employed at all level.
- *Evaluation:* Evaluation is the study of the degree of success, we get in terms of our original objectives of the programme.
- *Extension Agent:* The extension worker at the village level in direct day to day contact with local farmers.

- *Folk Ways:* Folk ways are customary ways of behaving in which society exerts some what, they are not rigid.
- *Functional Literacy:* Functional literacy is the ability to read, write and calculate, so that a person may engage in all.
- *Farming Need:* Farming needs are the gap between farming practices adopted by the farmers at present and the farming practices recommended by the research worker at present.
- *Field Trip:* Field trip is a group of interested farmers and extension education workers who go on tour to see and obtain first hand knowledge of improved practices.
- *Flow Charts:* Flow chart shows the functional relationship, *e.g.*, organisational set up in a block, district or state.
- *Film Strips:* These are the series of still picture in one roll.
- *Folk Songs:* Folk songs are a powerful means of propagating the idea and experience of a wise and the learned.
- *Flannel Graph:* It is a piece of flannel or sand paper stick on flannel.
- *Forum:* Two or more speakers who present talks on the same subject. The primary characteristics of the forum is that the subject is controversial and speakers present the opposite sides of questions.
- *Family:* Family is a system of organised relationship involving workable and dependable ways of meeting basic social needs. It is a most multifunctional of all institutions in society.
- *Flash Cards:* These are small cards of thick paper with pictures, diagrams or drawings on a particular subject or topic.

- *Films:* Films are a series of motion pictures projected in sequences with a synchronised sound system.
- *Group Meeting:* Group meeting are a device to save time by several individuals simultaneously.
- *Goal:* Goal is the distance in any given direction proposed to be covered in a given period of time.
- *Graph:* Graph is a diagrammatic presentation of a quantitative data in a clear, interesting and easily understandable form and is very effective in showing trends, developments and relationships.
- *Individual Meeting:* These are between two or three individuals arranged at home or at the field.
- *Integrated Rural Development:* IRD is defined as development which can help to increase the purchasing power of the rural poor through the generation of greater opportunities of gainful employment.
- *Innovation:* Innovation is an idea perceived as new by the individual.
- *Innovativeness:* It is the degree to which an individual is relatively earlier in adopting new ideas than the other members of his social system.
- *Interest:* Interest refers to the attention with a sense of concern focussed on some objects.
- *Illiteracy:* Illiteracy means inability to read and write.
- *Illiterate:* It means person who has not acquired basic skills in reading, writing or arithmetic.
- *Institution:* Institution is a system of relationships or a pattern for carrying out an idea or a desire which is regarded as necessary for the welfare of the group.
- *Learning:* Learning is the process by which an individual through his own activities becomes changed in his behaviour.

- *Lecture Note:* Lecture note is merely a store of information.
- *Lesson Plan:* Lesson plan is an outline of activities that the teacher will follow in order to create effective learning situation for the learners. It is a plan for teaching.
- *Leadership:* Leadership is a process of influencing the behaviour of the individual in a real situation.
- *Leaflets:* Leaflet is a single sheet of printed matter folded or unfolded giving informations on a particular topic.
- *Leader:* An individual is a leader in any social situation in which his ideas and actions influence the thoughts and behaviour of others.
- *Learning Experience:* It is a mental/physical reaction, one makes through seeing, hearing or doing the things to be learnt through which one gains meaning and understandings in solving new problems.
- *Method Demonstration:* Method demonstration are those in which a new method, or technique is demonstrated, taught or given practice of.
- *Motivation:* Motivation is the process of creating a strong desire for appealing to the fundamental wants of the people so that they are influenced to change the behaviour.
- *Models:* It is the replicase of real objects or sealed representation of things.
- *Mass Meeting:* These are the meetings of a general mass people called to hear a leader or endorse the decisions of an executive committee formed by the general body of an organisation.
- *Monograph:* Monographs are a technical publication, written on a particular subject.
- *Message:* It is the information, a communicator wishes

his audience to receive, understand, accept and act upon.

- *Newspapers:* Newspapers are an effective medium for disseminating information about local, district, provincial, country and world affairs.
- *Need:* Need is a difference between two situations that "what is" and what ought to be."
- *Norms:* Norms refers to the rules that govern action directed towards achieving values.
- *Objective:* Objective is the direction of movement.
- *Overtime:* Over time refers to the time required to adopt from its origin.
- *Opinion Leaders:* Those individuals from whom others seek the formation and advise.
- *Organisation:* Organisation is a set of two or more parts which are so related that they function as a whole.
- *Plan of Work:* It is an outline of activities so arranged as to enable effective exe^ution of the entire programme.
- *Project:* Project refers to an outline of procedure and portions to some phase of extension work.
- *Personal Example:* It is one of the most effective ways of training and communicating with the people. It is generally what we do and how we do it that makes a lasting impression on trainer or people.
- *Programme:* Programme is a systematic statement of situation, objectives, problems and solutions. It is relatively permanent but requires constant revision.
- *Pamphlets:* Pamphlets are very valuable and effective written material or literature for use in extension.
- *Programme Planning:* It is a process of working people in an effort to recognise the problems and determine possible solutions.

- *Planning:* Planning is a process of preparing systematic statement of the line of action for achieving certain overall and specific objectives in relation to needs and resources.
- *Panchayati Raj:* Panchayati raj refers to a members or an assembly of elders who settles disputes, according to custom in a village or within a caste.
- *Poster:* Poster is an important visual aid helping to arouse people.
- *Psychology:* It refers to the use of animals in behavioural research and collect information and in sight relevant to their area of concern.
- *Panel:* Panel is also termed as round table discussion. It is usually composed of 4-6 persons. The members of the panel carry on a discussion of the topic among themselves and the audience listens in on the discussion.
- *Poverty:* Poverty is the social phenomena in which a section of the society is unstable to fulfil even its basic necessities of life.
- *Personality:* Personality is a pattern body of inherited traits.
- *Prestige:* Prestige is the evaluation of the status.
- *Result Demonstration:* These are those in which two practices or techniques are compared for results.
- *Rural Development:* Rural development is a process of change form the traditional way of living to progressive way of living.
- *Reference Group:* Reference group is a group of persons whom an individual consult before taking an important decision.
- *Role:* Role refers to the function of the status.
- *Rural Sociology:* Rural Sociology is the systematic study of the people living in rural areas and follow patterns

of occupation and life somewhat different than those living in urban areas.

- *Staff Meeting:* These meetings are among formal groups staff meeting occupy a unique position. They are an essential part of a successful extension programme.
- *Subject Matter Specialist:* Professional staff who provide technical back up to extensionist.
- *Social Days:* Social days are days on which the extension and the administrative personnel of the block and the district get together with the villagers in some village.
- *Social Problems:* It is a condition in society that is considered harmful or undesirable by society as a whole based on existing social values.
- *Social Centre:* The adult education classes will be organised through different adult education centres on the basis of needs and requirements of adult men and women.
- *Social Class:* Social class is an abstract category of persons arranged in levels according to the social status they posses.
- *Social Group:* Social group is a unit of two or more persons in reciprocal interaction or communication with each other.
- *Social Value:* These are defined as the attitudes held by individuals, group or society as a whole as to whether material or non-material objects are good, bad, desirable or undesirable.
- *Social Stratification:* Social stratification is the pattern of super imposed categories of differential privileges.
- *Social Sanction:* Social sanction is a reward or punishment applied to persons in order to encourage or discourage certain types of behaviour.
- *Social Welfare:* It refers to a large number of activities

connected with the uplift and welfare of the weak and exploited sections of the society.

- *Social System:* A population of individuals who are functionally differentiated and engaged in collective problems and solving the problems.
- *Status:* It is a position or place in a social set-up.
- *Social Control:* Social control is the process of constraining persons to conform to various norms.
- *Sociology:* Sociology is the scientific study of men's behaviour in relation to other groups and individuals with whom he interacts.
- *Specimen:* It may be a part of an object or one of group or class to represent the whole group.
- *Supervision:* It is the process of placing into action standards of performance developing and utilising human resources to achieve organisation results.
- *Society:* Society refers to the group within which men share.
- *Television:* Television is one of the most powerful audio visual aids. It combines in itself both visual and auditory aids and is very effective in conveying actual life events in action.
- *Traditions:* The uniformly accepted ways of thinking.
- *Time Charts:* They present data in ordinary sequence, *e.g.*, year to year increase in the income of the people, profit and loss account of farms, etc.
- *Technical Workers:* These workers are trained in special fields. Their training includes some basic training in specialised fields followed by job training.
- *Teaching:* Teaching is the process of arranging situations that stimulate and guide learning activities in order to bring desired changes in human behaviour.

- *Visual Aid:* These are aids which facilitate understanding of a problem or topic through seeing.
- *Village Level Worker:* The multipurpose agent of India's community development programme at village level.
- *Wall-newspapers:* These are big sheets of paper with current news, experiences, recommendations, etc., printed on them.
- *Walking Tours:* Walking tours are an important means of communicating with people.

Main Principles

The principles discussed below one by one:

(1) Principles of interest and needs.
(2) Grass-roots principle of organisation.
(3) Principle of cultural differences.
(4) Principle of cultural change.
(5) Principle of Cooperation and participation.
(6) Principle of applied science and democratic approach.
(7) Principle of learning by doing.
(8) Adaptability principle in the use of extension teaching methods.
(9) Principle of trained specialists.
(10) Principle of leadership.
(11) Whole family principle.
(12) Principles of satisfaction.

Extension by Diverse Agencies and Organisation

(a) Work at Shantiniketan: In 1908, Shri Rabindra Nath Tagore, under his scheme of rural development work, started youth organisations in the villages in the Kaligram pargana of his zamindari.

- In 1921, he established a rural reconstruction institute at Shantiniketan.

Objectives: (1) To create a real interest in people for rural improvement. (2) To study rural problems. (3) To improve village sanitation.

(b) Gurgaon Experiment: Rural welfare movement on a mass scale was first started by Mr. F L. Brayne, Deputy commissioner in the Gurgaon district of the Punjab, adjacent to Delhi.

- Under his programme, a village guide was posted in each village who was to act as a channel through which the advice of experts in various departments could be passed on to the villagers.

(c) Marthandam Programme: Marthandam programme was set up under the guidance of Y.M.C.A. (young men's christian association) in Travancore.

- The leader in this work was Dr. Spencer Hatch, an American agricultural expert. The essential technique of the centre was 'self help with intimate expert counsel.'
- From the demonstration centre at Marthandam, about hundred villages were covered through Y.M.C.A centres in villages.

(d) Etawah Pilot Project: The idea of starting this project was conceived and born in 1947, it was put into action with headquarters at Mahewa village about 17 miles form Etawah (U.P.) in September 1948.

- Mr. Albert Mayer of U.S.A. who came to India with the American forces in 1944 was the originator of the project.

He started it with the aim of introducing intensive work on the rural reconstruction front.

(e) Indian Village Service (I.V.S.): Its founder was Mr. Arther T. Mosher of New York and Shri B.N. Gupta who established it in 1945.

- The objectives of this organisation were to assist village people to realise the best in their own villages by developing individuals, volunteer leaders and local agencies and enabling them to be effective in helping themselves and others.

(f) Sarvodaya Programme: It was a Gandhian concept and evoked great enthusiasm in Bombay state.

- The main features were simplicity non-violence, holiness of labour and reconstruction of human values.
- Its aimed at raising the standard of living, scientific development of agriculture, promotion of cottage industries, spread of literacy, medical and health facilities and the development of village panchayat.

(g) Firka Development Scheme of Madras State: It was Government sponsored and aimed adoptment of the Gandhian ideals of Gram Swaraj by bringing about not only educational, economic, sanitary and other improvements in villages but also by making the people self-confident.

- The scheme was launched in the last quarter of 1946 in 34 Firkas throughout the state and on April 1, 1950, it was extended to another 50 additional Firkas.
- The collector, who was primarily responsible for the successful working of the scheme in the district.

(h) Nilokheri Experiment: It was originally started to set up again 7000 displaced persons from Pakistan and later integrated with the 100 surrounding villages into what came to be a rural cum urban township.

- It was built round the vocational training centre that was transferred from Kurukshetra in July, 1948 to the 1100 acres of swampy land on the Delhi-Ambala highway.

(i) Intensive Agricultural Development Programme (I.A.D.P): The intensive agricultural development programme was popularly known as a package programme. This name was given because of the collective and simultaneous application of all improved practices.

- These practices were improved seeds irrigation, fertilizers, plant protection, implements, credit, etc.
- The programme was started in July 1960 in 7 selected districts in various states.

These were:

(1) West Godavari (A.P.)
(2) Shahabad (Bihar)
(3) Tanjore (Madras)
(4) Raipur (M.P.)
(5) Ludhiana (Punjab)
(6) Pali (Rajasthan)
(7) Aligarh (U.P.)

Role in Progress

Extension education centres around three units

(a) Farmers or rural peoples

(b) Innovations or inventions

(c) Extension workers

- The technological development of education can be related to various fields of extension like Agriculture and Home Science.

Agricultural Extension: a Theory

Agricultural extension is concerned with agricultural education assisting farmers to bring about continuous improvement in their physical, economic and social well being through their individual and Cooperative efforts.

It makes available to the farming community the scientific and other factual information the training and guidance for the solutions of the problems in agriculture including Animal Husbandry, Gardening, Agricultural Engineering, etc.

- Role of Agricultural Extension: Each country has to develop its extension organisation according to its own social cultural, economic situations at a given stage in its development. There are however, certain broad generalisation which can made it. These are

(1) Extension work can be most effective only when it is entirely educational in nature and free from regulatory and other govt. responsibilities.

(2) As development proceeds, more specialisation at the field level may be feasible and desirable.

(3) The successful work of the extension officer at the field level depends upon their being supported by specialists at the next higher level of administration.

(4) Efforts should be made to keep extension and research closely related to each other.

(5) Extension work must be directly towards the family as a whole men, women and children.

- Work with youth and farm women is important.

(6) In the interest of providing maximum effective service within the limited resources, there is an urgent need for effective coordination of all the agencies serving rural people (IFS 2001).

Progressive Approach

Among the communication and educational techniques 'demonstration' are the oldest, best and simplest tools for transmitting sophisticated technology in a simple and understandable form.

- Demonstration means showing by doing. The basic principles of demonstration is learning by seeing and doing. There are three types of demonstration

Method Demonstration: It is the short-type demonstration. The method demonstration evolved by Dr. Seeman. A. Knapp. So the father of method demonstration is Dr. Seeman. A. Knapp. This is a single practice demonstration, *.i.e.,* how to apply fertilizer in the field. This cannot compare between old and new technique or skill.

There are some points which should be taken into consideration while demonstrating practice.

- *Phase of Demonstration:* An effective demonstration programme involves 5 essential phases:
  (1) The first phase is of analysing the situation.
  (2) The second phase is deciding on goals to be accomplished in conducting a demonstration.
  (3) The third phase is the demonstration plan, and execution of demonstration.

  The plan includes the three stages:
  (i) Pre-demonstration stage
  (ii) Execution stage
  (iii) Follow-up or post-demonstration stage.

  (4) The fourth phase is evaluation of the demonstration programme.
  (5) The fifth phase is reconsidering after evaluation has taken place.
- Important tips conducting a good demonstration
  (1) The demonstrator should encourage the audience and see that people understand and learn to do what is being demonstrated.
  (2) Before starting the actual working, he should tell the audience.
  (3) Tell the audience especially the interested farmers.
  (4) Help the audience to go through the steps of demonstration so that they can repeat the process themselves without further help.
  (5) A demonstrator should be well-prepared in the subject himself and should take the help of his colleagues and teachers, whenever necessary.
  (6) Invite questions and create an atmosphere that the audience may clarify it doubts.

Result Demonstration: In this demonstration, the result of demonstrated practices are shown in comparison to existing practices:

- In result demonstration process, the extension worker may take the help of audio-visual aid like film, slides, pictures and comparative diagrams, etc. So the farmers imagination is fired and they have some idea of how things were done to obtain higher yields.

Significant Points for Result Demonstration

(1) The students should demonstrate an specific problems of the area-drilling wheat against broadcast, hybrid maize against local variety.

(2) It should be carried out correctly, in detail and must be in sequence.

(3) Better to conduct more than one demonstration in the same village, on the same subject.

(4) Never try to discover new truths.

(5) Do not repeat demonstration repeatedly.

(6) Use audio-visual aids to support the result of demonstration.

(7) The demonstration should be located by a well-travelled roadside or path.

Visual Aids: Visual method of communication is more effective than other. Out of the five physical senses through which we learn, the eye is the most effective (helpful) in learning.

Planning for use of Visuals: For the use of our visuals we should think of objective, audience, media, resource and technique.

- The subject matter of a visual is picked up after establishing the objectives of the visual.

Visual Aids-non Projected: The non-projected visual aids are divided into two groups: (A) Teaching aids (B) Display visuals.

(A) *Teaching Aids:* Teaching aids are non projected visual aid and can be prepared and used by teachers or extension workers are:

(1) Chalk board

(2) Pictures, including photographs and hand drawn pictures

(3) Flash cards

(4) Flip books—similar to flash cards

(5) Lip charts—few charts on a particular theme put together and bound together at the top.

(6) Flannel graph.

(B) *Display Visuals:* Displays visuals is not projected aid. It is also non-projected visual aid. Its examples given below:

(1) *Posters:* A poster helps the extension workers to get across one idea to the audience. It is a good communication method.

(2) *Diagrams:* We can draw a picture of any speciman like goat, cow, building plan, etc., it is called diagram.

(3) *Charts:* A chart contains a series of idea. Generally poster is used in the awareness stage but a chart is used in the testing stage. Chart can be

(a) Pictorial chart (b) Organisational chart

(c) Flow chart (d) Suspense chart

(e) Stream chart

(4) *Graphs:* For getting a precise idea or drawing a comparison or statistical data needs presentation of idea through graphs. Different types of graphs found there

(i) Line graphs (ii) Tree or stream graphs

(iii) Area graphs (iv) Bar graphs

(v) Pie graphs (vi) Photographs

(5) *Flip Chart:* It is like an album of drawings, pictures, charts. It is an assists an extension workers or teachers to tell story.

(6) *Flash Cards:* These are the series of cards which when presented before the audience in proper sequence, tell a complete story.

Each card is of about 10" to 12" in size and contain a picture or diagram.

(7) *Felt Board:* This is a board with a surface of felt, flannel or khaddar cloth which is used for fixing pictures or parts of material on the back of which there is piece of sand paper to stick

(a) *Chalk Board:* Even it is called blackboard. This is helpful in meeting and group discussions. This is generally used in villages and other places.

(b) Bulletin Board: It is a board with a background of coloured cloth. It can be covered with glass, netting or a plane board. The material for display may be news sheets, booklets, cartoons, circular letters.

- This is a good medium for displaying the creative work of the students.

(c) Exhibitions: Exhibition is one of the best programme of extension which the extension workers can arrange successfully. An exhibition can show actual things, practices, programmes in progress through charts, diagrams, displays, models, etc.

Visual Aids being Projected

Such aids include slides, film-strips and films. The projection of these aids involves various types of equipment beginning from the simple magic lantern to the complicated sound fi m projectors.

Types of projection: (1) Direct Projection, (2) Indirect Projection, (3) Reflected Projection Projectors

(1) Opaque Projectors: In India and England, opaque projector is also called as Epidiascope.

(2) Slide Projectors: Slide projectors are based on principle of direct projection.

(3) Overhead Projector: This equipment uses the principle of indirect projection.

(4) Movie Projectors: Movie projector may be classified on the basis of size of film.

Sound Recording of Audio-aids

There are three methods of recording the sound:

(1) Mechanical

(2) Magnetic process

(3) Optical process

Disc recording is done by the mechanical process.

- Tape and wire recording is done by the magnetic process.
- Recording of sound on a movie film by the optical process.

Disc Recording: In the discs or gramophone records, discs of the lac and plastic are usually used. The rate of revolution per minute of the disc are 45, 78 or $33\frac{1}{2}$. The $33\frac{1}{2}$ r.p.m. records are long playing having 16 inches diameter and taking 15 minutes for each side.

Wire Recording: The wire employed for this purpose is a steel wire which has diameter of 4-5 mils. The process of magnetic recording on this wire is based on the working principle of an electro-magnet.

- This wire is available in spools which are of 3 types:

(1) lasting for 15 minutes

(2) $\frac{1}{2}$ hour

(3) one hour

Tape Recording: Using the principle of magnetic process recording can be done on tapes made of paper, plastic or metal.

The tape recorder accommodate two spools, one empty take-up spool and other containing tape for recording.

Radio: Radio is the good source of communication of ideas to the rural people especially in a country, like India. It is the good source of extension of agricultural information to farmers. The All India Radio stations publish a Journal Akashvani which gives the metres and frequency in kilo cycles on medium and short waves.

Non-projection of Audio-visuals

Puppets: Puppets can be very interesting aids for telling stories to the village people the puppets may be:

(1) Hand puppets, (2) String puppets,

(3) Rod puppets, (4) Shadow puppets

Extension Education and Drama

This is a good source of entertainment and extension education for the villagers. The students of the extension wings of agricultural with the help of their teachers, teachers of village schools, VLW's and villagers can arrange to produce dramas in their villages.

Talks: Even talk a good source of communication for this purpose should be good speaker. Confidence and courage are necessary for making a good speech.

Meetings: For this process of communication three stages given below:

(A) Planning

(1) Selection of topic

(2) Timing

(3) Selection of place

(4) Selection of speaker and chairman

(5) Advertising and publicity

(6) Physical arrangements

(B) Conducting the Meeting

(1) Programme procedure

(2) Audience participation

(C) Follow-up of Meeting

(1) Summaries

(2) Press reports

(3) Displays

(4) Other method

(5) Evaluation

Kinds of Meeting

(1) Lecture forum

(2) Panel discussion

(3) Symposium

(4) Group interview

(5) Public conversation

(6) Debate discussion

(7) Workshop

JOINT CONFERENCES

We can participate in conferences more effectively by keeping the following tips into consideration

(1) Tell the audience in advance how many points you intend to make and specify when you come to each of these points.

(2) Be confident that you can do a good job.

(3) Speak naturally in a loud, clear but not a shouting voice.

(4) If possible, put an outline of points on a blackboard or chart.

(5) If the situation permits, speak to some of the audience members personally.

(6) Use only brief references and notes if possible avoid reading it.

(7) Take every possible opportunity to speak.

(8) Avoid giving a talk right after a meal. Mental capacity is at lowest after eating.

(9) Give an eye-catching title to your talk.

Explanation of Tours

A tour is essentially a visit to demonstrations, plots, farms of progressive farmers, dairy, model house, etc.

Objective of tour

(1) To see the results of a new practice.

(2) To see a new practice demonstrated.

(3) To see the operation of a new tools.

(4) To see the accomplishment of other village, college, farms, etc.

Communication through Written Words: It is playing a great role in communication with farm people, home, markets.

- Following methods for written communication:

  (1) Newspaper (2) Stories

  (3) Magazine articles (4) Leaflets

  (5) Pamphlets (6) Circular letters

  (7) Wall news papers

- Leaflets contains 4-12 pages stiched or stapled at the centre.
- Pamphlet varies in size form 12 to 24 pages.
- A bulletin is a bulkier publication with 24-48 or more pages.
- Circular letter is a quick and cheap method of communication and proves very effective if done well.
- Circular letters can teach villagers easily.

Planning of Future Events

Programme planning is a procedure of working with the people in an effort to recognise unsatisfactory situations and to determine possible solutions or objectives and goals.

Kelsey & Hearne (1949):

> *"An extension programme is a statement of situation, objectives, problems and solutions."*

Lawrence 1965:

> *"An extension programme is the sum total of all the activities and understanding of a country extension service. It includes*

(1) The programme planning process

(2) Written programme statements

(3) Plan of work

(4) Programme execution

(5) Results

(6) Evaluation

Components of Programme Planning

The components of programme planning are complex. There are:

- People their needs
- Useful technology
- The educational process
- Analysing situation
- Making decisions about what should be done
- Determining action
- Projection the desired shape of things into the future, etc. Principles of Programme Planning

For the development of a useful and effective programme, the following principles should be kept in mind, which is suggested by Kelsey and Hearne (1955).

(1) Extension programme planning is based on an analysis of the facts of the situation.

(2) Extension programme planning select problems based on the needs and interests of the local people.

(3) Extension programme planning determines priorities; defines objectives and solutions which offer satisfaction.

(4) Extension programme planning has performance with flexibility.

(5) Extension programme planning balance with emphasis.

(6) Extension programme planning has definite plan of work.

(7) Extension programme planning is an educational process.

(8) Extension programme planning is a continuous process.

(9) Extension programme planning is a coordinating process.

(10) Extension programme planning involves local people and their institutions.

(11) Extension programme planning provides for evaluation of results.

India Based National Plan

Basic Plans: These plans which were formulated before independence were

(a) *The Bombay Plan:* This plan was called " A plan of economic development of India." It was prepared by eight businessmen of Bombay in 1945. The main objective of this plan was to double up the per capita income within the period of 15 years. It was divided into 3 stages of 5 year each.

(b) *The People's Plan:* It was prepared by Shri M.N. Roy in 1944. Its guiding principle was " Production for use as distinct from production for exchange".

(c) *Visvesvaraya Plan:* This plan was prepared by Shri M. Visvesvaraya and accepted by the All-India manufacturers organisation in 1944. Its primary object was to raise the country's economic efficiency to a level at which the average citizen would be able to find employment and to earn a livelihood.

(d) *Gandhian Plan:* It was prepared by prof. S.N. Aggarwal, based on Gandhian ideas. Its objective was to raise the material as well as the cultural level of the masses to a basic standard of living within a period of 10 years.

Establishment of National Planning Committee: A national planning committee was appointed by the Congress working committee of India in 1938, with Shri Jawaharlal Nehru as its chairperson. The committee laid out certain guidelines so as to realise planned advancement from year to year.

Planning Commission: The planning commission is a multi-member body. The Prime Minister of India is the chairman of the planning commission. The Minister of Finance is the member in charge of finance in the commission in an ex-officio capacity. These arrangement ensure co-ordination between the commission and the central ministries.

Schemes for Rural Development

Various rural development programmes were started after in the year of 1960.

(a) Intensive Agricultural District Programme (IADP): The intensive agricultural district programme, popularly known as the package programme, was started on a pilot basis in 1961 in 7 selected districts. The programme goals at combining technical know how credit and production supplies for stepping up agricultural production.

- The operation of IADP has contributed to a significant improvement in the use of critical inputs like improved seeds, fertilisers and plant protection measures.

(b) *High Yielding Varieties Programme (HYVP):* The high yielding varieties programme was started in 1966-67. The cultivation of HYVP has resulted increase in food grains production. Wheat production has been more than doubled. Rice production has also increased through not as much as in the case of wheat.

- A major development in rice cultivation was the launching of the Minikit programme from 1971-72.

(c) *Drought Prone Areas Programme (DPAP):* The drought

prone areas programme formerly known as rural works programme was initiated in 1070-71 as a non central sector scheme. The main aimed at mitigating the severity of scarcity conditions by executing rural works to generate employment.

(d) *Integrated Rural Development Programme (IRDP):* Integrated Rural Development Programme is a non-plan programme which means IRDP comes under five year plan.

- The main aims of IRDP is the all round development of family.
- IRDP was launched on 2nd Oct. 1980 by integrating following programmes

(i) *SFDA*—Small Farmers Development Agency

(ii) *DPAP*—Drought Prone Area Programme

(iii) Operation Flood II

(iv) *NREP*—National Rural Employment Programme

(v) *MNP*—Minimum Need Programme

District Rural Development Agency

The DRDA (District Rural Development Agency) is headed by the collector/ deputy commissioner or CEO zila parishad in states.

- DRDA helpful for Implementing National Rural Employment Programme (NREP). National Agricultural Research Project

NARP was launched in January, 1979 and is founded by the ICAR and through IBRD. The soft loan connected of the world bank.

- The NARP is the first major effort of the Indian Council of Agricultural Research (ICAR) to promote and strengthen the regional research capability of the State Agricultural Universities (SAUs).
- The project is designed to fill the gaps in research.

NAEP: NAEP (National Agricultural Extension Project) is being launched by the Government of India. So that on effective and efficient transfer of technology is assured. The links establised between NARP, IRD, and T and V system together produced the desired results.

- NAEP is being launched with the help of IBRD- World Bank connected to overcome the various constrains identified earlier and will rectify many of them.

T and V System: Training and Visit System is a new extension service approach. The concept of T and V system was evolved in 1973 by Israeli extension expert Daniel Benor. Therefore, T and V system is also called Baster and Benor scheme and firstly evolved in Turkey (in Seyhan irrigation project).

- T and V system has two phages/stages: (i) Training, (ii) Visit
  - (i) Training: Training is imparted for transfer of technical know how from subject matter specialist to extension worker.
  - (ii) Visit: Extension worker has to visit to the farmers for transfer of know how obtained at the training.
- T and V programme is Govt. sponsored rural development programme.
- T and V system is introduced in India in 1974 in Rajasthan canal area and chambal command area in M.P. to improve the effectiveness of agricultural extension.
- Training and visit programme also called Intensive Agricultural Extension.

Various Programmes

Lab to land programme was launched in India on the occasion of ICAR golden jubilee celebration year in July 1979 for only one year. But later on this programme was extended further year after year.

- That aims of lab to land programme is the transfer of proved and viable low cost agricultural technologies to small and marginal farmers and landless laboures to improve their overall socio-economic condition.
- That's why lab to land programme was launched under lab to land or land to lab programme, the problems related to the agriculture were brought to the laboratory where the particular solution was to be found out or searched out and later on were communicated to the farmer.

Krishi Vigyan Kendra (KVK): Education Committee (1964-66) recommended to establish agricultural polytechnique. At last ICAR mooted the idea to establish Krishi Vigyan Kendra (KVK) and Mohan Singh Mehta Committee was appointed in 1974 by ICAR to formulate KVK scheme.

- First KVK was started at Pondichery in 1974 with the aim of transfer of technical literacy to increase agricultural production.
- The main purpose of KVK is to impart training at the Kendra or out of the centre.
- There were 314 KVK in number by the year March 2001 and plan is to establish one KVK in every district of India.

The credit for the success of KVK goes to Dr. Chandrika Prasad.

Establishment of Community Development Programme: In community Development programme, chief role was played by grow more food campaign. On 2nd Oct. 1952, 55 Community Development Project were started in different parts of the country with Indo USA agreement for three years. This project covered nearly 25,260 villages and a population of 6.4 millions.

- The basic idea and concept of community development was borrowed from USA.

Panchayati Raj System: Beginning

Panchayati Raj was first started at Nagaur Rajasthan and

Andhra Pradesh in 1959. As a whole state Rajasthan was covered first by the Panchayati Raj system.

- In the Panchayati Raj system, village Panchayat is the lowest unit. The General body of Panchayat is Gram Sabha.

TRYSEM Scheme: Training of Rural Youth for self employment. TRYSEM was launched in 1979 with the basic principle of knowledge by doing or learning by doing to provide technical skill to the rural youth from families below the poverty line (BPL)

- The target group comprises rural youth between the age of 18 and 35 years old. (Ins, PCS, 2000, 2001)

Establishment of 4H-Club: 4H-Club was established in 1900 in America to provide training to youth for agricultural and home development, it was first govt. step.

- 4H-Club means Heart, Head, Hand and Health was explained by A.B. Graham.
- On the basis of 4H-Club of USA, young farmer's association of India was started by P.S. Deshmukh in 1956.

Important Abbreviations

IRDP — Integrated Rural Development Programme
TTC — Trainer's Training Centre
AU — Agricultural Universities
CB — Cooperative Bank
FTC — Farmers Training Centre
WVDP — Whole Village Development Programme
KVK — Krishi Vigyan Kendra
(ICAR, RAS 1999; JRF 2001; ARS 2001; U.P.PC.S. 2000)
TAD — Tribal Area Development
HAD — Hill Area Development Programme

MFAL — Marginal Farmers and Agricultural Labour Programme

IDAD — Integrated Dryland Agricultural Development

DPAP — Drought Prone Area Programme

SFDA — Small Farmers Development Agency *(PCT 2003)*

CAD — Command Area Development

CDD — Cattle and Dairy Development

SD — Sheep Development

PD — Poultry Development

IADP — Intensive Agricultural Development Programme *(U.P. PCS 2002, IAS)*

IAAP — Intensive Agricultural Area Programme

HYVP — High Yielding Varieties Programme

IODP — Intensive Oil Seed Development Programme

ICDP — Intensive Cotton Development Programme

NCA — National Commission on Agriculture

ICRISAT — International Crop Research Institute for Semi-Arid and Tropics

LDB — Land Development Bank

ARDC — Agriculture Refinance and Development Corporation

JRY — Jawahar Rozgar Yojna

FCI — Food Corporation of India

NAFED — National Agriculture Cooperative Marketing Federation

IFFCO — Indian Farmers Fertilizer Cooperative Ltd.

AGMARK — Agricultural Marketing

RRB — Regional Rural Banks

FSS — Farmers Service Society

SBI — State Bank of India

IFCI	—	Industrial Finance Corporation of India
UTI	—	Unit Trust of India
IDBI	—	Industrial Development Bank of India
LBS	—	Lead Bank Scheme
WTO	—	World Trade Organisation
IMF	—	International Monetory Fund
NABARD	—	National Bank for Agriculture and Rural Development
NCDP	—	National Cooperative Development Corporation
AMA	—	Agricultural Marketing Advisor
APC	—	Agriculture Price Commission
NSC	—	National Seed Corporation
GNP	—	Gross National Product
TRYSEM	—	Training of Rural Youth for Self Employment
FAO	—	Food and Agricultural Organisation
CDP	—	Community Development Programme

2

Ecology and its Application

Significant Role

The mature community with its greater diversity, larger organic structure and balanced energy flow is often able to buffer the physical environment than the young community which is often the more productive. Thus from envolutionary point of view, the primary function of ecological succession is achievement of stability rather than a mere increase in productivity.

The principals of ecological succession are of great importance to mankind. For man, it is necessary to have early successional stages as a source of food since he requires a large net primary production to harvest. On the other hand, the stability of the climax and its ability to buffer and control physical factors are important from the human population point of view. The only way man can have both a productive and a stable environment is to insure that a good mixture of early and mature successional stages are maintained with interchanges of energy and materials. Excess food produced in young communities helps feed older stages that in return supply regenerated nutrients and help buffer extremes of weather.

In the most stable and productive natural situation there is usually such a combination of successional stages. For example, favourable situation exists in many terrestrial landscapes. Here productive croplands on the plains are intermingled with diverse forests and orchards on the hills and mountains. The crop fields are young stages maintained by the farmer. The forest represents older, more diverse and self-sustaining communities with lower rate of production. If the forests are destroyed, water and soil may wash down from the slopes and reduce the productivity of the plains. Thus for our own safety and protection, some definite per cent of the landscape should be occupied by more or less natural communities, *i.e.* forests.

Definition of Ecology

KNIGHT (1965) has defined ecological succession as an orderly sequence of different communities over a period of time in some particular area. According to SMITH (1966) ecological succession is an orderly and progressive replacement of one community by another till the development of a stable community in that area. BENTON and SMITH (1969) have proposed another definition of ecological succession according to which succession is the occurrence of gradual, orderly and predictable changes in the composition of communities towards the climax type.

From the above definition it can be inferred that the development of community in an ecosystem begins with pioneer stage which are replaced by a series of more nature communities until a relatively stable community is formed which is in equilibrium with the local conditions. The whole series of communities which develop in a given area is called sere; the relatively transitory communities are called seral stanges or pioneer stages and the final stabilised community is called the climax community.

Succession Procedures

The principle of ecological succession is one of the most important aspect of ecology. The remarkable thing about it is that it is directional and predictable; and results from the modification of the physical environment by the community. In the end it culminates in a stabilised ecosystem in which maximum biomass and symbiotic function between organisms are maintained. The progressive developments in an ecological succession can be enumerated as follows:

1. There is usually progressive development of the soil with the increasing depth, increasing organic content and increasing differentiation of layers towards the mature soil of the final community.
2. The height, massiveness and differentiation into strata of the plant community increases.
3. The rate of organic matter formation per unit area in the community increases with increasing development of the soil and of community structure.
4. As density of above ground plants increases, the microclimate with the community is increasingly determined by characteristics of the community itself.
5. Species diversity increases from the simple communities of early succession to the richer communities of late succession.
6. Population of pioneer stages rise and fall and replace one another with the time gradient due to interspecific and intraspecific competition for space and nutrients. This interaction and reaction among plants species and between environment provide a continuing course of succession through the modification of environmental factors such as soil, moisture and humus. The rate of this replacement shows through the course of succession as smaller and short lived pioneer species are replaced by larger and longer-lived ones.
7. Consequently, the relative stability of the communities

increases and the final community which is usually stable is called climax. The climax community is dominated by longer lived plants which maintain their population in equilibrium with community composition.

Kinds of Ecological Succession: Ecological succession may be of the following two types:

Initial Succession

When succession begins on an area which has not been previously occupied by a community, *e.g.* a new exposed rock area, sand dunes, new islands, deltas, shore or recent lava flow, it is known as primary succession. The first group of organisms (plants or animals) which become established in such an area is termed the pioneer community.

Later Succession

When community development is proceeding in an area from which a community was removed and where nutrients and conditions for existence are already favourable, *e.g.* cutover forests, abandoned cropland and ploughed field, it is termed secondary succession. Secondary succession is usually more rapid because some organisms are already present. Moreover, previously occupied territory is more receptive to community development than the sterile areas.

Depending upon the predominance of green plants of heterotrophic organisms in the initial seral stages, succession is distinguished into autotrophic succession and heterotrophic succession.

1. *Autotrophic Succession:* It is wide spread in nature and begins in a predominantly inorganic environment. It is characterised by early and continued dominance by autotrophic organisms (*i.e.* green plants).
2. *Heterotrophic Succession:* It is characterised by early dominance by heterotrophs such as fungi, bacteria and animals. It occurs in a predominantly organic

environment such as in a stream heavily polluted with sewage, in fallen log or in small pool receiving leaf litter in heavy quantities. Energy is maximum at the beginning and declines as succession occurs unless additional organic matter is imported or an autotrophic community takes over. But in autrotrophic succession, energy flow does not usually decline but is maintained or increased.

Normally succession or the replacement of one type of community by another is due to the modification of the environment by the communities themselves. This type of succession is known as autogenic succession. In some cases the replacement of one type of community by the other is due to the forces other than the effects of communities on the environment as in ponds and streams where nutrients and pollutants enter from outside and modify the environment. This is known as allogenic succession

Phenomenon of Succession

In either primary or secondary succession, a pioneer is the first seral stage to migrate and become established in a bare area from the surrounding areas. The aggregate organisms of such a community are called the pioneer species and such community is known as pioneer community. The pioneers react on the soil and modify the environment to make successful hold of the soil. To be able to tolerate the adverse climatic and edaphic conditions that often exist in a new area, the pioneer species should have the following characteristics:

1. The population of pioneer species should be small in size so that the nutrients requirements is low.
2. It should be able to survive on low nutrient level.
3. It should have the capacity to bring nutrients in solution to be absorbed later on.
4. The rate of increase in number and growth should be as low as possible.
5. It should be more dynamic to face interspecific and intraspecific competition.

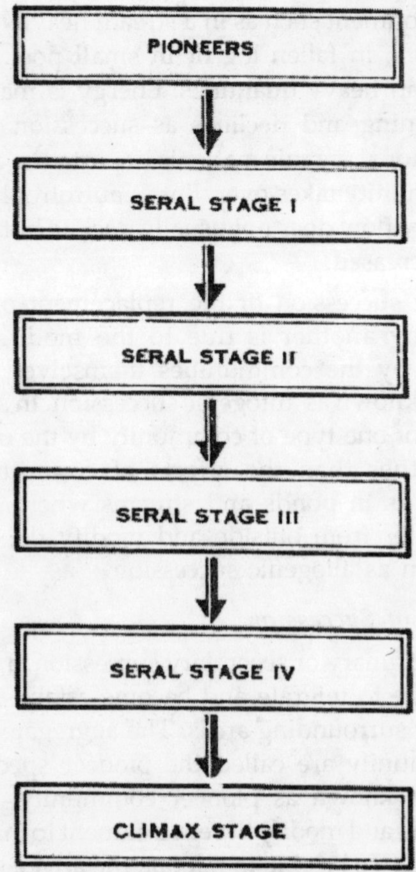

Hypothetical representation of the process of ecological succession.

The pioneer community generally exists for a short interval of time, in comparison with some of the later successional stages. Due to the increase in their number, there starts a competition among the different organisms of the community for food and shelter.

To meet their requirements of food and for shelter, the organisms react on the environment and modify it. Such changes in environment lead to increased tension and the gradual inability of the pioneer species to cope with the altered conditions. As this situation continues, other species infiltrate into the area because the environment becomes more suitable for their existence.

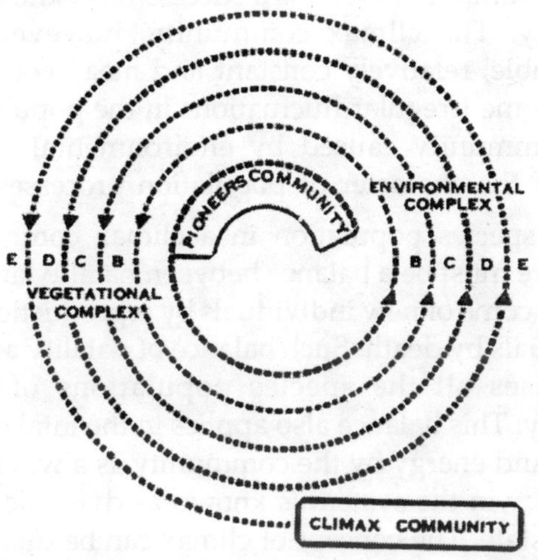

Diagrammatic representation of the process of succession with different plant communities appearing under the influence of developing environmental complex taking hydrosere as an example.

By the death and decay of pioneer species, the organic matter, moisture and nitrogen contents increase considerably in the soil. The enriched soil is more suitable for the growth of secondary communities which become established and the pioneer species gradually disappear. Each secondary community represents a seral stage. Each seral stage appears, flourishes and finally disappears as the environmental changes occur. Each seral stage has its particular community called the seral community. The various seral stages of ecological succession leading to a particular group of climax are referred as a sere.

When a climax community has become, established, it tends to remain in possession of the area because it does not change the environment so as to injured itself or to favour the growth of different dominant species, and because its members can resist the competition from the organisms entering from outside.

Climax Principles

The community that ends a succession is known as climax community. The climax community, however, fluctuates around stable, relatively constant and mean conditions. This is due to some irregular fluctuations in the populations of the climax community caused by environmental changes and because of the character of population processes.

For a species population in a climax community to be stable, there must be a balance between natality and mortality, between income of new individuals by reproduction and outgo of individuals by death. Such balance of natality and mortality characterises all the species populations of the climax community. This balance also applies to the intake and release of matter and energy by the community as a whole. This kind of constancy in the system is known as dynamic equilibrium or steady state. The concept of climax can be understood best by taking a pool along a stream as an example. Water flows through the pool, but with equal rates of inflow and outflow, the pool itself as a system stays the same, *i.e.* in steady state. Similarly climax implies a steady state or condition in a natural community.

Concepts of Succession

Two theories concerning climax communities have been advanced by ecologist namely the monoclimax theory and the polyclimax theory.

1. *Monoclimax Theory:* This theory was put forward by an American ecologist FREDERICK CLEMENTS in the year 1916. According to this theory there is only one climax community in a given climatic or geographical region. But due to topographic irregularities and various soil type within that climate zone, there are other communities also in that zone in addition to climax community. These communities are known as subordinate communities the terms such as proclimax, postclimax, disclimax, preclimax and subclimax are used to define those communities.

Proclimax is refereed to that community which is more or less stable resembling the climax community in some respects. When disturbed by man or other animals, it is called disclimax community such as grassland in forest area. When the development is arrested in the subfinal stage of succession due to repeated burning, cutting or grazing and flooding, etc., such stage is termed subclimax or preclimax community. Sometimes, a strip of vegetation of higher life from is found within climax as a forest strip along a stream in a grassland. This higher type of vegetation is called the postclima community.

2. *Polyclimax Theory:* A number of European ecologists (NICHOLOS, CHAMPION and BORE) along with WHITTAKER (1953) advocated this theory. According to this theory a number of different climax communities may be present in a climatic or geographical region. Thus preclimax, postclimax and disclimax communities are all climax communities, as they are all self-reproducing and are able to maintain for an indefinite period of time. This theory does not rule out climax communities but the other stages within the area are also given equal recognition and are not regarded as minor communities.

3. *Climax Pattern Theory* (WHITTAKER 1963; MCINTOSH 1958 and SELLECH 1960): According to this theory, the composition and other characters of the climax communities are determined by the total environment of the ecosystem and not by a single factor. Thus the climax community may be represented by a pattern of animal communities which have close inter-relationship with the flora and fauna of the area, the soil and climate. A particular plant cannot thrive unless its seeds are properly distributed or a certain animal cannot live, if food plants are not available. Among the most popular animals and plants are the climatic climax which is the community which expresses the climate of the area concerned. Among the present day ecologists the theory seems to be popular.

Styles of Succession

Depending upon the types of habitat and varying amount of moisture, the successions are variously designated. The chief patterns are hydrosere or hydrarch (in water), xerosere or xerarch (in dry conditions) and mesarch or mesosere (an intermediate type with adequate moisture.

Role of Xerosere

One of the best examples of xerosere is the succession which starts on bare rock, wind blown sand, rocky talus slopes and such place where there is extreme deficiency of water. The various stages in xerosere can be enumerated as follows:

1. *Lichen Stage:* Due to great exposure to sun and extreme deficiency of water, the first pioneers on the bare rock area are a few simple organisms. The most successful of such organisms are crustose lichens. These are able to withstand extreme desiccation due to excessive dryness. During rainy season they absorb large quantities of water and flourish rapidly. Migration to distant rocks takes place either by spores or soredia by wind. The common species of crustose lichens are Rhizocarpon, Rinsodina, etc. These begin the long, slow process of rock disintegration. Rock particles and dead organic matter of lichens accumulate to provide conditions possible for the growth of higher forms of lichens.

   As soon as little soil is formed by the activity of crustose-lichens, higher forms of lichens such as foliose-lichens appear. These include Dermatocarpon, Parmelia, Umbilicana, etc. These have large leafy thalli which overlap the crustose lichens and cause their gradual death and decay. In this way more and more humus accumulates and gradually a thin layer of soil is formed which consists of rock particles, remains of lichens, dust particles and moisture.

   Associated with the lichens a few mites make their

appearance. Along with them, a few spiders and ants also make their appearance in cracks and crevices of the rock.

2. *Moss Stage:* With the accumulation of dust and humus in small quantities, the environment is altered enough to allow the establishment of secondary communities in a rather definite sequence. Scattered patches of mosses such as Tortula, Grimmia, Beryum and Barfula, etc. begin to invade the environment that had so far been dominated by lichens. Later on, mosses like Funaria, Sphagnum and Polytrichum make their appearance.

   Among the animals, mites become more varied; some small spiders and spring tails as well as tardigrades become associated with this secondary community.

3. *Herbaceous Stage:* As the mats of mosses become more extensive more soil accumulates; much of the soil is blown in from surrounding areas during windy periods. More mineral material is added to the soil as acids leach out from the overlying vegetation and increase the depth of the mineral soil layer. Many annual weeds develop which are, later on, followed by biennial and finally perennial grasses. Andropogon, commonly known as broom sedge, becomes a dominant grass in many areas.

   With the influx of grasses, the fauna (animals) also becomes varied. Nematodes and larval insects, collembola, ants and mites appear in the gradually altered environment.

4. *Shrub Stage:* Further modification of the environment provides conditions for the germination and growth of shrubs and perennial woody plants such as Acacia, Prosopis, Capparis, Zizyphus, etc.

   With the approach of shrubs, the animals also become vivid and numerous and join hands with the vegetation in altering the environment.

5. *Climax Forest:* With the establishment of shrubs, more and more soil is formed and environment becomes increasingly humid. This favours the growth of woody trees. In the beginning trees show stunted growth and are sparsely placed. Finally a climax forest community is established and a number of terrestrial vetebrates make their appearance. The climax community is the last aggregation in the successional series. If the climax conditions do not change and no catastrophic event alters the area, the community maintains itself indefinitely.

Functions of Hydrosere

Hydrosere or hydrarch succession starts in water. A freshly built pond can be taken as a most suitable example of hydrarch succession.

The various stages of hydrosere can be enumerated as follows:

1. *Submerged Stage:* In initial stages water is poor in nutrients and devoid of life. The pioneers in an aquatic habitat are plankton. The phytoplankton grow floating or suspended in water and multiply. With the death of phytoplankton and zooplankton, the substratum is enriched with organic matter. As water becomes rich in organic and mineral substances, certain rooted submerged hydrophytes make their appearance.

   Prominent among them are Ceratophyllum, Potamogeton, Vallisneria and Utricularia, etc. By the death and decay of these plants there is further enrichment of the medium. With the increase in nutrients, the level of the pond is raised and it becomes shallow.

2. *Floating Stage:* When water level in the pond remains only 6 to 7 feet, deep, floating plants begin to appear. These plants include Nymphaea, Nelumbium, Trapa

and Monocharia, etc. These have their roots rooted in the mud and their leaves freely floating at the surface. Later on free floating plants like Lemna, Azolla and Wolffia grow profusely to cover the water surface.

3. *Reed-swamp Stage:* As the water body becomes shallow by silting and removal of water by transpiration, the environment becomes less suitable for the free floating and submerged plants. Now reed-swamp plants like Typha, Rumex, and Sagittaria invade the area. The reed-swamp plants build up the shores by retaining the sediment and accumulation of plants remains. Beavers, muskrats and other animals carry material into the pond, deciduous vegetation blown in from the shore, and silt is carried in from the surrounding land. Rafts of vegetation from the pond margin drift offshore, strand, and take root, thus establishing islets that grow in size until they meet also and join the shore.
4. *Marsh-meadow Stage:* As the free water is changed to swampy land, the water plants give way to swampy plants such as sedges and rushes. As succession continues marshy meadows become too dry for swampy plants and these are subsequently replaced by herbs and shrubs.
5. *Woodland Stage:* As succession continues, the soil is further built up so that it becomes drier and is also changed chemically. In time certain smaller species of trees invade the area, taking the place of the shrubs, and eventually full size forest trees will dominate the scene. At the same time when the vegetation is undergoing these profound changes in the hydrarch succession, the animal life of the community is corresponding altered. Fish, beavers and muskrats are gradually excluded and land vertebrates make their appearance.

As discussed there is a progressive change in plant life starting from pioneers (green algae) to the climax stage (forest) during hydrosere. Accordingly, the animal life changes as follows:

> Protozoans like Paramecium, Euglena, Amoeba, etc. which form the plankton, are the pioneers. When planktonic growth becomes rich, other animal forms like bluegil fish, sunfish (Lepomis), large mouth bass (Micropterus) make their appearance. Small caddis-flies which feed microorganisms also come to occupy the habitat.

Later on, when the higher forms of vegetation make their appearance, caddis-flies are replaced by other species that are able to creep over submerged vegetation and make cases from plant material. Such forms are represented by dragon flies, mayflies and crustaceans like Gammarus, Daphnia, Cyclops, Cypris, etc.

At the floating stage when floating vegetation makes its appearance in a pond or lake, the animal life is chiefly represented by Hydra species, gill breathing snails, frogs, diving beetles, and other insects. Besides these, larval forms of a large number of insects, water-snakes and turtles also come to occupy the habitat.

At the reed-swamp stage, the pond becomes shallower, the organic content of water increases, bottom starts exposing and the floating plants are replaced by emergent plants. With this, floating animals are replaced by those which can live in association with the emergent plants.

Different species of mayflies and dragon flies make their appearance on the upper surface of emergent plants while their nymphs remain attached to the submerged parts of the vegetation. Gill breathing snails replaced by lung breathers as Lymnea, Physa and Gyrulus.

Among insects, water scorpion (Ranatra), scavanger bettle (Hydrophilis) and giant water bug (Lithocercus) make their appearance. The bottom of the pond is inhabitated by certain annelids and sunfish is replaced by bull head and mudpickrels.

Great blue heron, king fisher, red winged black birds, swamp sparrow, ducks, beavers and muskrats, etc. become common to the area. At the sedge meadow stage, the pond my become dry during summer. Therefore, fish beavers and muskrats are gradually excluded. In their place snails like Psidium and Anodonta become common. Finally, at the woodland stage, most of the terrestrial animals come to occupy the habitat.

Functions of Merosere

It is an intermediate type with moisture present in adequate amounts. The successional series is much shorter because moisture conditions are more ideal, and the initial water problems that must be resolved in a xerarch or hydrarch type of succession are non-existent. In actuality, the conditions towards which hydrarch and xerarch communities are gradually progressing are those that prevail in the mesarch series, thus giving the later type of sequential pattern a head start.

Necessary Elements

Ecology is an interdisciplinary branch of biology which deals with the study of distribution structure, various aspects of life of organisms and their interaction with the environment.

The French Zoologist, Isodore Geoffrey St. Hilaire proposed the term ethology for "the study of the relations of the organisms within the family and society in the aggregate and in the community."

St. George Jackson Mivart coined the term hexicology, which he defined as "devoted to study of the relations which exist between the organisms and their environment as regard the nature of the locality the frequent, the temperatures and the amount of light which suit them, and their relations to other organisms as enemies, rivals or accidental and evolutionary benefactors."

General Description

The term ecology (Oekologie) was first used by Reiter in 1885 followed by Ernst Haeckel in 1886. It was coined by combining two Greek words oikos (meaning 'house' or 'dwelling place') and logos (meaning the study of) to denote such relationships between the organisms and their environment.

Thus, literally, ecology is the study of organisms 'at home'. Misra has defined ecology as the study of interaction of form, function and factors. Krebs described ecology as "the scientific study of interactions that determine the distribution and abundance of organisms."

Ecology differs from other branches of biology firstly it comprehends physical environment alongwith organisms and secondly it deals with level of biological organisation. The various levels of biological organisation are called biological spectrum.

Level of organisation	*Basic division of biology*
Organic molecules	Subcellular organelles
Cells	Cell Biology
Tissues	Histogy
Organs	
Organ Systems Organisms	Anatomy, Physiology
Biochemistry	
Population	
Communities	Ecosystems Ecology

However, the recent development in study of ecology has been the recognition of the fact that the biotic (living) and abiotic (non-living) components of the nature are not only interrelated but both these components function in an orderly manner as a definite system, thus presently it is considered to be the structure and function of ecosystems. Ecology is a synthetic branch of science which obtains source materials from several other disciplines like climatology, geography, geology, limnology, mathematics, physics, chemistry, microbiology besides other branches of biology.

It is because ecology is to deal with the study of interactions of living beings among themselves as well as their physical environment. The physical environment consists of soil, water, air and energy. It provides both habitation and raw materials for the synthesis of organic food.

Several other life processes are dependent upon the physical environment, *viz.*, germination of seeds, pollination, respiration, hibernation, etc. like components of living world, the various constituents of physical environment are interrelated. Some of their interactions are useful to living organisms, *e.g.*, ozone, nitrogen and carbon dioxide filter ultraviolet rays from incoming solar radiations while infra red rays are absorbed by the water vapours present in the atmosphere.

Each level of organisation has properties peculiar to it that are not identifiable at the levels below. But studies at higher levels must take account of the lower levels. Ecology has three levels of organisation such as population, communities and ecosystems. However, a strict demarcation between lower and ecological levels cannot be made since the latter also involve the study of effects on the constituents of lower levels of organisation.

Population: A population is a group of individual organisms of the same species (in breeding individuals) in a given area. It is also called local population. A local population adapted to its particular environment is called ecotype. A species may have several ecotypes.

The different ecotypes have different genetic set-ups within the broad frameworks of the gene pool of the species. They are capable of interbreeding. Where two ecotypes meet, a continuous change from one type to another occurs due to interbreeding. It is called ecotine.

Community: A community is a group of population of different species in a given area. It may include all the population in that area—all plants, all animals and microorganisms or may be defined more narrowly as a particular group such as the fern community, or the seed-eating bi-community of that area. The different population do not remain secluded but show interactions and interdependence.

Vegetation: A vegetation is a group of plants growing under the same climate over a wide range.

Ecosystem: An ecosystem is a segment of nature consisting of a community of living organisms and the physical environment, both interacting with each other and exchanging materials between them. Ecosystem ecology emphasises the movements of energy and nutrients among the biotic and abiotic components of the ecosystems.

Because the ecosystem is the highest level of biological organisation, all ecological concepts can be set within its framework. The bioti:: components of any ecosystem are linked as foodchains. Food chains are interlinked to form complex food webs. Food webs are the basic units of ecosystem ecology. Thus ecology begins with population and culminates in ecosystem.

Biosphere: The different ecosystems of the world are not independent rather exchange material among themselves. The biologically inhabited part of the earth consisting of all the ecosystems of the world is called biosphere.

The Manners

Like other sciences ecology too has its own principles and basic concepts. Earlier, stress was laid down on describing structural components and their relationships but later on emphasis was laid on the functional aspects of various components. There is also a trend to study historical or evolutionary causes of ecological phenomena.

Fundamental Requirements

1. The different types of organisms living in a particular environment are not only independent and mutually reactive but affecting each other in various ways. Animal population, Flora and vegetation are interdependent through the environment and are mutually reactive.
2. Environment, which is actually a complex of several interrelated and dynamic (variable) factors some of which become critical at critical stages of the life cycle of the species. It, therefore, works as a sieve selecting organisms that can flourish in it.
3. The species put each effort to maintain its uniformity in structure, function, reproduction, growth and development by preservation of its genetic pool. However species have sufficient plasticity to modify themselves according to changing environment by virtue of osmatic plasticity (ecads) or by recognition of their genes during sexual reproduction (ecotypes).
4. It is not only the environment which influences the life of organisms, but organisms too modify their environment to make it more congenial for their growth, development, reproduction and dispersal. The modified environment may become less suitable for the community already living in it. The dynamic environment and organisms make ways for the development of different kinds of organisms through a process known as succession. The process continues till the development of a community which is now more or less stable and is now able to keep itself adjusted in equilibrium with the environment. This final stage of community is called a climax.
5. According to Clements and Shelford (1939), under similar climatic conditions, there may develop more than one communities, some reaching to climax stage, others under different stages of succession. This complex of many communities growing in a particular area, sharing a common climate is called a biome.

Operational Factors

1. The basic structural and functional units of nature are ecosystems. An ecosystem is a segment of nature consists of community of living organisms and physical environment both interacting and exchanging materials between them.
2. Discrete biological unit consists of a number of population each occupying a specific functional position with respect to other organisms with which it interacts. It is called ecological niche.
3. The degree of success of a particular population in an area is determined by the parameters of both abiotic factors as well as interaction with other types of population. These interactions among organisms can be positive, negative or even neutral.
4. Also, there are involved energetics of ecosystem, as energy is the driving force of this system. The flow of energy in the ecosystem is unidirectional or non-cyclic. The radiant energy is trapped by the autotrophic organisms (producers) and is transferred as organic molecules to the heterotrophic organisms (consumers) and decomposers. Energy is lost during its transfer from one trophic level to the next. Organisms use the energy in respiration.
5. A number of inorganic nutrients are taken up by living organisms for their metabolism and body building and are called biogenetic nurtrients. These chemical components keep on moving between biotic and abiotic components of an ecosystem in defined cycles called biogeochemical cycles.
6. Human beings have exploited the ecosphere to an extent that only economically important plants are allowed to grow in an ecosystem. This caused a reduction in species diversity and natural interaction amongst organisms and physical environment.

7. Under natural conditions, different population undergo succession. Ecosystem undergo an orderly process of change with time, passing from a less complex to a more complex state. This process involves not only changes in species but also changes in physical environment of a community. The terminal or stabilised state is known as climax.

Growth Factors

1. Present day organisms are the product of evolution of the past organisms.
2. By knowing the evolutionary aspects of each ecological phenomena, we can predict the future shape of events under different sets of conditions.

Various Mechanical Skills

Ecology is an interdisciplinary approach for studying environment. It is considered both science and art. It concerns both the aspects doing and understanding. An ecologist has to work with living system and various environmental components. In ecology various types of tools and techniques are used. It employ scientific and non-scientific methods, tools and techniques. An ecologist has to begin with observations of living organisms and environmental components. The observations are quantified and statistical techniques are used to analyse the relationships and interdependence. The observation tool and correlation techniques an commonly used by ecologists the qualitative and quantitative method: and tools are employed in ecology.

Qualitative Methods	*Quantitative Methods*
(i) Observations	(i) Observation
(ii) Historical	(ii) Experimental
(iii) Longitudinal	(iii) Cross-sectional
(iv) Descriptive	(iv) Project method
(v) Sampling Technique	(v) Sampling technique

The qualitative is the subjective approach while qualitative is objective and scientific approach.

Atmosphere and Ecosystem

The term "Environment" which etymologically means surroundings, is considered as a composite term for the conditions in which organisms live and thus consists of air, water, food and sunlight which are the basic needs for all living beings and plant life, to carry on their life functions. The environment also includes other non-living things, temperature, wind, electricity, etc. In other words, environment consists of both biotic and abiotic substances. Environment creates favourable conditions for the existence and development of living organisms. Environment can be defined in number of ways:

(1) "Environment is the sum of all social, economical, biological, physical or chemical factors which constitute the surroundings of man, who is both creator and moulder of his environment."

(2) "Environment refers to the sum total of conditions which surround man at a given point in space and time."

(3) "Environment is the representative of physical components of the earth where in man is the important factor influencing his environment."

"Environment" is assuming great importance in our dairy life. Though environment is a very broad term but in common parlance it is the 'limited surround' that affects the functioning of the world in which we live, thus in our daily life, what we are really concerned about is the "Biosphere". The part of the biosphere which can be affected by the present day technology, is called the Technosphere' and in the part of 'Ecosystem' that is badly affected.

It is very difficult to maintain ecological balance until and unless poverty elimination programmes are pursued effectively. Our then Prime Minister in his speech on 6th July 1987 in the meeting of the World Commission on Environment and Development emphasised that there should be global commitment for mass poverty elimination. If we are able to improve the quality of life of the poor, we will be able to protect and regenerate the environment.

After the scientific and industrial revolution in the recent past, there has been immense impact of man on his environment. Man has failed to realise that any new factor upsets the balance of the ecosystem as a whole. All the devastating effects of man's effect to control nature have occurred because he has upset the balance relationship of the organisms that make up the environment.

Huge industrial installations every year, introduction of faster mode of transport and sprouting up of large crowded cities (Urbanisation) are the main outcomes of the modern civilization. These and a large number of many others are contributing to what is called environmental pollution. Increasing industrialisation is also causing much danger to man's life because it is also responsible for defouling or polluting of environment by man.

Mankind will perish if the protection of the environment does not become an integral part of all the development programmes because we cannot cheat the "Mother Nature".

Reliable on plant foods rather than animal foods is ecologically more viable and sustainable because of the tremendous loss of energy in the food chain of ecosystem. The conservation of natural resources especially forests and wild life got attention in both developed and developing countries. Late Sanjay Gandhi gave much attention to the preservation of trees find cultivating more of them.

The Government of India, along with other Governments participated in a United Nations sponsored conference on the human environment in June 1972 at Stockholm (Sweden). The conference concluded that the decline in environmental quality is due to increasing pollution, loss of vegetable cover and biological diversity and excessive concentration of harmful chemical in the environment.

In May, 1986 Government of India passed the Environment Protection Bill which afterward became an act. The act gave a wide meaning to the environment and concluded that the environment included water, air land and the living creatures such as plants, animals and human beings. To implement Environment Protection Act throughout India our Government launched National Environmental Awareness Campaign, organised All India Essay Competitions on environment for the school children, National Seminars on various aspects of the environment, established Environmental Education Centres (Ahmedabad).

The subject of environment got immense national priority. Its protection and promotion was made one of the twenty programmes of the Prime Minister. How the plant life and healthy environment affect our mental health is not even cleared to the well informed people. There is an equal need for protection to oil and petroleum by minimising its wastage.

It was categorically emphasised in an exhibition on "Our Environment and Our Future" held at International Trade Fair, New Delhi in 1989. We alongwith other developing countries, have to find alternative path to an alternative good. A goal which ultimately is the true goal of development—an environmentally sound and sustainable development. Time has reached when we are facing challenge to our intellect and wisdom for saving the humanity from extinction. I arises from our activities constricting the environment and depleting the biosphere, especially so in the name of development. We have to plan our actions ecologically for the sustenance of the environment and development. Perhaps the greatest challenge facing the present generation is need to balance apparently conflicting responsibilities of caring for the exploitation, preservation and conservation of environment around them.

Misra (1991) presented an excellent account of planning for environmental management. The four basic requirements of environmental management recognised by him are, (i) impact of human activities on the environment (ii) value system, (iii) plan and design for sustainable development, and (iv) environmental education.

The conservation of natural resources especially forests and wild life, mineral, metals and oil got attention of the Govt. in both developed and developing countries. The problem is however, grave in over populated, poor and educationally backward society in which India is placed. Population, poverty and environment are interrelated. India has often been described a rich land with poor people. The nexus between poverty and environmental degradation can hardly be overemphasised. The low level literacy, economic backwardness and over population are some of the important aspects of Third World States Culture. A nation can be called super power largely due to its natural resources. All the branches of science should be involved in promoting the rational use of natural resources environmental protection, its improvement and welfare of human society.

The organisms live on different surroundings of the earth which include biotic (living) and abiotic components which form the environment. The individual pass through various stages in the life cycle, *i.e.* from birth to death. They have capabilities to continue their race from generation to generation. With the death of old individual new one start their cycle of life. For the existence of all living organisms on the earth, they need the system which sustains life. The important system is the environment which covers all living and non-living systems. So it is necessary for every layman and literate person to know the meaning of the environment.

Building-up Structure

Environment can be conveniently divided into two categories:

(1) Natural Environment

(2) Man-made Environment

Natural Environment: It means the environment that comes into existence without interference of man. The natural environment system operates through self-regulating mechanism called homeostatic environment mechanism, *i.e.* any change in the natural ecosystem brought about by natural process is counter balanced by changes in other component of the environment.

Many components together coordinate to form a natural environment which helps in sustenance of life. The environment consists of mainly two components:

A. Abiotic or Non-living components

B. Biotic or Living Components.

Abiotic Components: They are those components which are not living but can support other living organisms. These are the basic components of Environment and when these components become unbalanced they will prove fatal for the living organisms. These are of three kinds:

1. *Inorganic Substances:* These include elements like Nitrogen, Calcium, phosphorous, water and gases like carbon dioxide and oxygen. These substances are required for the synthesis of organic substances and are called inorganic nutrients or biogenetic substances.
2. *Organic Substances:* They are substances dispersed from dead bodies of organisms or are derived from their excretions. Organic substances together called organic detritus. It produces humus which is essential for maintaining proper soil aeration, hydration and fertility. Humus contains nucleic acids, amino acids, proteins, lipids, carbohydrates, etc. These are again sent back to the environment after decomposition by decomposers.
3. *Physical Factors:* They include climatic, edaphic and topographic factors, *e.g.* light, temperature, humidity, wind, rain, soil, topography and background, etc.

(i) *Light:* It provides solar energy for photosynthesis.

Maximum light energy is available at equator. It decreases towards poles. In a tree more energy is available to upper leaves than the lower leaves. Their rate of photosynthesis is accordingly higher.

(ii) *Temperature:* On the basis of temperature range, there are four life zones - tropical (hot throughout the year), subtropical (hot during summer and cool during winter), temperate (pleasant summer and cold winter with occasional snowy and arctic or alpine (Snowy throughout except for brief summer).

(iii) *Rainfall:* Amount and periodicity of rainfall determine type of vegetation.

(iv) *Humidity:* Transpiration is inversely related to humidity. Humid areas have abundant dense plant growth with numerous animals and epiphytes.

(v) *Wind:* It controls weather, transpiration, pollination and dissemination of propagules.

(vi) *Soil:* Its texture, pH and fertility determine the type of vegetation and animal community depends on it.

In addition to these, other abiotic components are:

(a) *Lithosphere:* Outer most layer of the earth

(b) *Hydrosphere:* Part of the earth having water resources

(c) *Atmosphere:* It is a cover around the earth having a number of gases

Biotic Components: The living organisms form the biotic component of the environment. Every living organism in the Biosphere depends upon one another and these living organisms which exist in Biosphere forming the biotic community of the environment are as follows:

(a) *Producers or Autotrophs:* They are autotrophic plants which synthesise organic compounds from inorganic raw materials in the presence of energy usually derived from the sun. Organic compounds serve two purposes—body building and liberation of energy during the process of respiration. All other organisms depend upon the producers for getting their food and energy contained in it.

(b) *Consumers or Heterotroph:* They are also called phagotrophs. Consumers are animals which ingest other organisms or their particulate organic matter. They are of two main categories—herbivores and carnivores. The herbivores or primary consumers directly feed on plant matter, *e.g.*, rabbits, deer, cattle, goats, horses, etc.

Carnivores feed on the flesh of other animals. Secondary consumers or primary carnivores eat the herbivores, *e.g.*, frog, mollusc, small birds, etc. The secondary consumers, in turn are eaten by tertiary consumers and so on. The carnivores which are not further prayed upon are called top carnivores, *e.g.*, Lion.

Since organic matter is resynthesised by consumers, they are also called secondary producers.

(c) *Decomposers:* They include bacteria, actinomycetes and fungi. They decompose the complex compounds in the dead organic matter of plants and animals and again recycle the elements into the environment.

Man-made Environment: Man is the most powerful environmental agent, spearheaded by modern technologies capable of modifying the environment according to his needs to a great extent without taking into account to his needs to a great extent without taking into account its consequences. Increase in the scientific technologies which are the product of human brain is now deteriorating the environment. Urbanisation, industrialisation and commercialisation all have lead to more critical conditions.

The condition of villages in worse as there is no proper sewerage and sanitation system. Increase production leads to spoilage of the environment and alter the composition of natural products. The development of any country depends on its geographical and environmental conditions. From the ancient time man is continuously exploiting and struggling with nature.

Degradation of environment started with the development in the early man's life in the form of many inventions and discoveries. These resulted in the development of new tools and implements. He also made developments in other fields such as industry, nuclear and other technological developments which are causing degradation of environment.

Development: From the early time man is putting efforts for development such as agriculture, urbanisation, industrialisation, commercialisation, etc. All these developments have put stress on the environment because these required large spaces which forests have to be cut without taking into account its consequences.

Advancement in agriculture lead to use of more chemical fertilizers, pesticides, weedicides, insecticides without taking into consideration their ill effects. The urbanisation stresses on basic necessities like food, shelter clothing which directly influenced the forests. Industrialisation posed a serious threat to biotic and abiotic components of the environment. As more and more industries are being set up those are polluting the environment with their emission and discharge effluents.

Population Stress: As population is increasing day by day their basic demands are at the increase. The ecological balance of the earth is disturbing because the natural resources are limited to meet their demands. This increase in population causing destruction of the environment. Therefore, it is necessary to save our environment by sustainable development. The increase in population should be at a rate that the resources should be exploited in a way that they not depleted.

Present Scenario: As we have already depleted lot of our natural resources it is the need of the day we should have a control on the unlimited use of our resources so that we could improve our environment. For this proper planning and management of resources as well as the environmental impact assessment should be done before establishing any developmental activity.

Every person of the country should be aware of the consequences of changed environmental conditions. Non-governmental organisations also play important role in educating people and controlling environment. So it is important to safeguard our environment which in turn safeguards the living beings.

Constituents of Atmosphere

Plant ecology is divided into four branches—autecology, synecology, system ecology and applied ecology. A perusal of the developmental history of ecology reveals that ecological studies made from time to time are based upon three principles aspects—taxonomic affinities, habitat and level of organisation.

Autecology: This is also known as ecology of individuals where we study individual species or population in relation to its environment from the time of seed germination upto setting of fruits and seeds, *i.e.,* seedling growth, vegetative growth and reproductive growth. An applied branch of autecology is agronomy which studies the various aspects of crop plants.

Synecology: Under natural conditions, however, organisms-plants, animals, microbes, etc., live together as natural group affecting each other's life in several ways. Thus, more complex situations exist where the unit of study, instead of single organisms are groups of organisms known as community. Synecology deals with the study of plant communities, including the composition, organisation and development in relation to environment. It is divided into several branches according to the aspect under study:

Habitat Specialisation (Habitat Ecology): Organisms are influenced and modified by the habitat in which they live. Therefore, they can be studied in a group according to the habitat, *e.g.*, freshwater ecology, marine ecology, grassland ecology, forest ecology, desert ecology, etc.

Community Type (Community Ecology): In ecology no organism can be studied in seclusion because of the interdependence and interaction amongst various members of a community. Therefore, community is studied as a single entity, *e.g.*, forest ecology, grass land ecology.

Taxonomic Group: In early days of ecology, botanist and zoologists engaged themselves separately in the study of ecology of plants and animals respectively. This leads to the development of such sub-divisions as:

(i) Plant ecology, and

(ii) Animal ecology.

In each sub-division there are further specialised fields like ecology of pines, insect ecology, avian ecology, bacterial ecology and ecology of mosquitoes, etc. However, modern ecologists feel that the principles in the study of plants and animals are not only much similar, but these two great groups of organisms are inseparably interrelated with each other. Therefore, modern ecology studies inter-relations and interdependence of all organisms.

Levels of Organisation: Depending upon the level of organisation, synecology can be divided into many types:

(i) *Population Ecology:* The study of growth, structure and regulation of population of organisms.

(ii) *Community Ecology:* Study of interdependence and interaction amongst various members of community.

(iii) *Ecosystem Ecology:* It deals with interactions of living and non-living parts of nature besides the flow of energy, biochemical cycling, trophic levels, etc.

(iv) *Biome Ecology:* Study of different communities living in a particular climate or area.

(v) *Genecology:* It is the study of ecology in relation to genetics.

There are some specialised branches of Ecology.

(i) *Geographic Ecology:* The study of geographical distribution of plants and animals.

(ii) *Radiation Ecology:* The study of effects of radiation on environment and organisms.

(iii) *Sociology:* The study of ecology and ethology of mankind.

(iv) *Ethology:* The interpretation of animal behaviour under natural surroundings.

(v) *Space Ecology:* The effect of space on the growth, development, reproduction and behaviour of organisms and cycling of resources so as to sustain life for long periods in the space.

System Ecology: It deals the analysis and understanding of the structure and function of ecosystem. It employs mathematical models.

Applied Ecology: The study of the human needs and the applications of ecological concepts, *e.g.*, agriculture, forestry, land use, etc.

There are basic fundamental principles which govern the various aspects of organisms and components of environment. The following are the fundamental principles of ecology:

(i) All living organisms and physical environment are mutually reactive. The organisms interact among themselves and affect each other and physical environment.

(ii) The physical and biological processes follow the principles of uniformitarianism which states the same physical and biological processes, as the environment is influenced by human activity.

(iii) Natural hazards affect adversely the biological communities in general and in particular. When biological processes are associated with physical events, yet severe hazards are created.

(iv) Sustained life on earth is a characteristic of ecosystem, not of individual organisms or population.

(v) Environmental principles of holistic nature of natural environment which largely affect the biological communities in a biospheric ecosystem.

(vi) Ecosystem is a fundamental unit of ecological study because it comprises both biotic and abiotic components.

(vii) Ecosystem functions through the input of energy mainly solar radiation which is trapped by green plants and is used to prepare food through the process of photosynthesis. The solar radiation is the main driving force of the ecosystem.

(viii) The ecosystem productivity depends on:

(a) The availability of the amount of solar radiations to the primary producers.

(b) The efficiency of plants to convert solar energy into chemical energy.

(ix) Ecosystem instability results when an ecosystem becomes unable to adjust with environmental changes.

(x) Man being an active agent of environmental change, modifies the ecosystem through the exploitation of natural resources.

3

ESSENTIAL FEATURES

The terrestrial environment is the most complex type of all, for in terrestrial environment there is a great fluctuation and interaction of environmental factors than in an aquatic medium. The interaction of physical, climate and biotic factors produces a wide variety of ecologic conditions on the continents and islands.

CHARACTERISTICS OF CLIMATE

Tundra biome is characterised by severe winter with a short growing period of 60 days. Mean monthly temperature varies between extremes of –35° (minimum) and +13°C (maximum), the warmest months averaging below 10°C. The annual precipitation is less than 4 cm. Most of the rain comes during summer and early autumn. But humidity is high and evaporation is low.

The ground remains permanently frozen and only the top soil upto a depth of only a few centimeters thaws in summer. The permanently frozen deeper layer is known as permafrost (RAY 1951).

Freezing and thawing lead to the fragmentation of large boulders into small rocks, cause downward slumping of soil slopes and lead to several irregularities in the topography. The general moulding of landscape by frost action is called cryoplanation. This makes the soil unstable and limits the growth and kind of vegetation.

Characteristics of Vegetation

Tundra is characterised by complete absence of trees. Its vegetation consists of sedges, rushes, lichens, mosses, decumbent shrubs, flowering herbs and grasses. These exhibit some common features. The plants are generally small with stunted growth and compact structure, adapted to desiccation and mechanical abrasion from wind and snow. Most of the plants are perennial because germination of seeds is poor.

Fauna: Inspite of extremely severe environment, the tundra gleems with life. The common mammals are reindeer, caribou, musk ox, arctic fox, arctic hare, arctic shrew, arctic ground squirrel, Tundra vole, lemmings, gray wolf, grizzly bear and polar bear etc. A large number of migratory birds visit Tundra during summer. These are arctic loon, red throated loon, goose, crane, gray falcon, hawks, plovers, gulls, larks, raven, snow bunting, arctic tern, snowy owl and purple sand piper etc. Reptiles and birds are either absent or represented very rarely.

Invertebrate fauna is comparatively restricted in variety. Snails, spiders, mites are well represented. There are swarms of flies and mosquitoes, just a few species of Lepidoptera, Hymenoptera and Coleoptera, whereas species of Hemiptera, Homoptera, Orthoptera and Odonata are almost absent.

Key Role of Air

1. The air is the source of oxygen and carbon dioxide for the living biota in terrestrial habitat. But none of them acts as a limiting factor, because of their great abundance.

2. Air acts as a medium of support for terrestrial biota and is less buoyant than water. An organism on land needs a comparatively rigid skeleton to live and move, whereas in water an organism needs relatively little support.

Key Role of Climate

The most important aspects of climate are moisture, temperature and light.

1. *Moisture:* In sharp contrast to aquatic forms, the terrestrial organisms lack constant contact with water. As a result moisture is a major limiting factor on land. The moisture content of air and soil varies from complete saturation (equatorial region) to slight amount (desert) at different places and in different weathers. Terrestrial organisms face the problems of obtaining water and preventing its excessive loss by evaporation and perspiration. Evaporation of water from the bodies of organisms and from the earth surface increases with rise in the relative humidity.
2. *Temperature:* Terrestrial ecosystem is heated by the absorption of heat by earth surface and air because of its low specific heat in comparison to water. Thus heat is gained and lost far rapidly by land than by water. Therefore, fluctuations of temperature diurnally and seasonally are more pronounced on land than in water. It varies from far below freezing point (-68°C) to 60°C in some deserts.
3. *Light of Solar Rradiations:* Solar radiations include radiations of different wavelengths like solar spectrum and ultraviolet and infrared radiations.

Different wavelengths have different effects on organisms. Ultraviolet radiations in large concentrations are harmful to most animals. Green light of the solar spectrum is reflected by plants and the rate of photosynthesis is maximum in red light.

Some animals (fish, snails etc.) exhibit an increased growth rate in different wavelengths in the following order-green, red, white, yellow, blue and violet.

In general terrestrial organisms are exposed to much higher intensities of solar radiation than the aquatic organisms and have evolved protective exoskeleton in the form of scales, hair, feathers, cuticle or chitin.

Key Role of Soil

In terrestrial ecosystems the soil serves two major functions:

1. It provides support for living organisms and it is a source of all essential nutrients (nitrates, phosphates etc.) for plants. The chemical nature and physical texture of the soil, sand and rocks is particularly vital in the distribution of organisms in terrestrial environment. Different types of soil have different properties and affect the availability of nutrients to plants and thus control the productivity of ecosystem and the nature of biota.
2. The land unlike ocean, is not continuous. Several geographical barriers hinder the free movements of terrestrial organisms.

Functions of Gravity

1. Aquatic organisms counteract gravity by evolving various floatation and swimming mechanisms. But a land animal has to support its entire weight. The animals best adapted to terrestrial life have developed appendages in the form of legs or wings for rapid locomotion over the surface or through the air.
2. Fluid buoyance permits water-dwellers to attain huge size, whereas gravity limits the mass of land animals.

The terrestrial biotic communities living in different regions of the earth vary according to temperature, moisture, available mineral matters, light, other forms of life and other limiting ecological factors. This means that regional climates interact with their regional biota and substrata and this results in the organisation of distinct easily recognisable community units, called biomes. A biome is a large land community unit, which can be easily recognised by its characteristic flora and fauna.

In each biome the kind of climax vegetation is uniform. Thus the climax vegetation of the grassland biome is grass, conifers in Taiga, deciduous trees in deciduous forest biome and so on, but the particular species of climax plants may vary even in different parts of the same biome.

The kind of climax vegetation depends upon the physical environment. According to plant ecologists a biome is identical with the major plant formation but it includes the animals varieties also and the occurrence of animal varieties is determined by the interaction of climax vegetation and physical forces.

The principal biomes of the world are:

1. Tundra,
2. Forest,
3. Grassland,
4. Desert,
5. Mountain and
6. Caves.

Role of Tundra Biome: Tundra Biome lies in fat north between the Arctic ocean and polar ice caps in the north and the coniferous forests to the south. There are two tundra biomes, one in the Palearctic (Europe and Siberia) and the other one in Nearctic region (North America).

Settlement of Animals

The animals living in Tundra exhibit following adjustments to the severe climatic conditions:

(i) *Loss of Pigmentation:* The animals are mostly white or

light gray in colour. This helps them to conceal and to protect against enemy.

(ii) *Hibernation:* Most cold-blooded animals remain inactive during colder months. Invertebrates pass winter in larval or pupil stage resistant to freezing. Rotifers, dysticid and hydrophilid beetles remain frozen in ice for months and resume activity in winter.

(iii) *Insulation:* Warmblooded mammals and birds have thick fur or dense plumage and a thick layer of fat. These measures help in retaining body heat.

(iv) *Hiding or Roosting:* Voles and lemmings etc. escape winter by staying in their nests or tunnels under snow.

(v) *Migration:* Some mammals and majority of birds in winter migrate to southern portions of Tundra or other warmer regions.

Forests as a Part of Biome

Forest is a large uncultivated and uninhabited tract of land covered with trees of different height (3 meters-30 meters), the shrubs (below 3 meters) and the herbs. Relatively good rainfall, high humidity, optimum temperature and presence of sufficient space condition the development of forests.

Forests as a Part of Atmosphere

The forest environment is less variable. Its uppermost level (tree tops) forms a canopy, which itself is exposed to sunlight and wind fury, but it does not permit them to the ground. As a result:

(i) Only diffused light reaches the ground.

(ii) Moisture contents of the soil remain more or less stable and are maintained for longer periods because of absence of evaporation.

(iii) There is no loss of water by transpiration below the forest canopy, because the tree tops act as effective wind breakers.

(iv) The temperature variations are peculiar. Temperature is generally lower in summer and is higher in winter. Also the days are cooler and nights are warmer than the surroundings.

(v) The floor of forest is thickly carpeted with encrusting thalloid plants like lichens and mosses. Several herbs also grow at the ground level.

Classification of Forests

Great variations have been observed in the nature and abundance or sparseness of forest vegetation. On that basis following types of forest biomes have been recognised.

1. Northern coniferous forests
2. Temperature forests
3. Tropical forests
4. Subtropical forests.

Northern Coniferous Forests: Coniferous forests are met with in the Northern Hemisphere, restricted to northern areas of North America, Europe, parts of Japan and Siberia. In Southern Hemisphere it is found in Australia and southern parts of South America. In Northern Hemisphere, its extensive geographical belt south of Tundra is known as 'Taiga'.

Climate: Coniferous forests are found in cold regions with heavy rain falls, high humidity and strongly seasonal climate. It has long winter with fairly short summer of about 60-150 days mean monthly temperature varies between –30°C (winter) and +23°C (summer) and faces heavy snow fall. The soils are acidic and mineral deficient and poor in nutrient. Because of the absence of evaporation, the soluble essential nutrients (calcium, nitrogen and potassium) are leached down sometimes even beyond the reach of roots.

Vegetation: Abounding these forests are coniferous evergreen trees of pine, balsam, larch, spruce and fir. These are the chief source of timber. The understorey trees are uncommon because of low light penetration. Orchids and ericaceous shrubs like blue berry constitute the only understorey vegetation. Mosses and lichens are found in abundance.

***Fauna*:** The vertebrate fauna is represented by mammals like lynx, bears, wolves, red foxes, squirrels, porcupines, star-nosed moles, lemmings, mouse, hares, deers, several groups of large sized mammals like moose, elk, caribou, beavers and wolverine etc. The common birds of this region are grouse, cross-bills, jayas, woodpeckers, flycatchers, saw-whet owl, grosbeak, sparrow, raven, wren, thrushes, warblers, finches and pigeon-hawk etc. Hylids are the only amphibians of coniferous forest and reptiles are few.

In the foot hills of the forest zone of India are found the tiger, the elephant, the Sambar deer, the gaur, the chital and the swamp deer The invertebrate fauna is dominated by insects like wood-wasps, bark-bettles, pine sawflies and pine-geometrids. Snails, annelids and millipedes are scarce.

Functions of Temperate Forests

The temperate forests are characterised by moderate climate with seasonal variations. These are of three types:

(1) *Temperate Forests:* These are met within North America, Western Europe, Eastern Asia (China and Japan), Chile and parts of Australia. The temperate deciduous forests have hot summer (maximum temp. 20-27°) and cold winter (–12°C) with snowfall. The rain comes in summer and the average rainfall varies from 30-50 inches per year. The summer, autumn, winter, spring and rainy seasons are distinctly merged. These seasonal changes have greatly effected the vegetation.

The deciduous forests are characterised by broad leaf trees, which attain a moderate height of 40-50 meters and form a thick canopy during summer. But during winter, the leaves are shed to withstand the excessive cold and drought. The principal trees are pines, oaks, hickory, maple, cidar, cotton wood, basswood, beech, chestnut, elms and willows. The understorey of shrubs and herbs is well developed, but herbs are more numerous. Often there are two blooms of herbs, the spring flower bloom immediately after winter before the trees have expanded their leaves, and the rainy season bloom which grows under low light level. There are a few vines (the grape) and some epiphytes, mosses, algae and lichens growing on tree trunks.

The invertebrates are represented by the earthworms, millipedes, snails, land isopods, wood-boring larvae of various insects and the leaf eating insects. The mammals are represented by squirrels, deer, bears, gray foxes, bobcats, pigs, dormice and ground dwelling wild boar, wolves, racoons, badger and shrew, goat, antelope, wild cat, lynx etc.

The flying mammals include bats, flying squirrels, arboreal porcupines and opossums. The avian fauna is represented by hawks, owls, woodpeckers, thrushes, crested flycatcher, horned owl and a good variety of crows, ravens, herons and storks, etc. The reptiles are represented by turtles, timber rattle snakes and amphibians by newts, salamanders, toads, tree frogs and cricket frogs etc. The deciduous forests are dominated by ground dwelling and cursorial forms.

(2) *Temperate Evergreen Forest:* This forest is characterised by warm dry summers and cool, moist winters. It is represented in North America (chaparrel of Pacific Coast), maquis in Mediterranean region, encinar in Spain and melle scrab on Australia's south coast.

In evergreen woodland the trees are absent, but shrubs range upto 3-4 meters in height. The forms are intermediate between the temperate deciduous forest and the drier grassland biomes. The plants possess great power of regeneration and sprout very quickly after being burnt by forest fire.

Animals characteristic of this woodland are wood rats, chipmunks, brush rabbits, mule, deer, lizards, wem-tits and brown towhees. The mammals are mostly cursorial ungulates, and jumping or saltatory forms.

(3) *Temperate Rain Forest:* The temperate rain-forest has a definite seasonality. The temperature as well as rainfall vary throughout the year. Rainfall is high but often fog is very thick and serves as an important source of water. Therefore, the temperate rain forest is the coldest.

Due to availability of water throughout the year, the diversity in flora and fauna is less. The dominant trees are redwood of Pacific Coast of North America; alpine ash of Australia and Tasmania. These are very high measuring even more than 100 meters in height. Epiphytes and lianas are common but not so rich as in any other rain-forests. Fauna is similar to other temperate forests.

Functions of Tropical Forests

Tropical Forests are found in tropical zone of the world and are characterised by very high temperature with rainfall in abundance. These are of two types:

(i) *Tropical Rainforests:* Tropical rain forests occur near the equator in Central and South America, Central and Western Africa, Southeast Asia (Malaya and parts of India, New Guinea, Java, Sumatra, Malaya, Borneo) and Northeast Australia. These forests are characterised by minimal seasonal changes. These have high and even temperature throughout the year. The maximum temperature ranges between 32°C-40°C and minimum between 10°C-0°C. The days and nights are of uniform length. The annual rainfall varies from 250-400 cm (100-160 inches) and generally it is uniformly distributed throughout the year. Due to high temperature, there is speedy evaporation of water from the heated land surfaces. When these vapours rise up these are cooled and result in heavy down-pour accompanied by lightening and thunder mostly in the afternoon. The relative humidity is always very high and seldom drops below 60 percent of saturation.

Soils in the tropical rain forests are red latosols and are exceedingly thick. Leaching of minerals at a very fast speed makes these soils unfit for agriculture. But there is speedy cycling of nutrients which makes these soils very fertile.

The flora of tropical rain forest is very rich and highly diversified. One square mile of land may contain 300 different species of trees. The trees are very high *i.e.* 36 to 56 meters (100-180 fit) and rich in thick stemmed woody lianas.

The tropical forests have a very rich fauna both in density as well as in varieties. The invertebrate and cold blooded vertebrates have attained the largest possible size. The common invertebrates are planarians, worms,snails, millipedes, centipedes, scorpions, spiders, isopods and insects. Among insects bugs, beetles, ants, bees, flies, termites, mantids, phasmids and some other orthopterans like basalattids are very numerous. Land leeches are found on the floor of forest as well as on the bushes.

The vertebrate fauna is very different in its different subregions and is dominated by arboreal and nocturnal forms. The common mammals found in forests are tiger, panther, leopard, lion, jackal, elephant, deers, sambar deer, chital deer, bison, tapirs, hippopotamus, scaly anteaters, wild pig, jungle cat, giant flying squirrel, flying phalangers, monkeys, howling monkey, spider monkey, black langur, gorillas.

The birds are represented by a large number of varieties and display vivid colours. These are woodpeckers, fruit eaters, humming birds, parrots, hornbills etc. The reptiles of tropical forest are geckos, chameleons, agamid lizards and many species of poisonous and nonpoisonous snakes and some aquatic reptiles. Frogs and toads are very common and arboreal amphibians are represented by Rhacophorus.

(ii) *Tropical Seasonal (=deciduous) Forests:* These are found in south-east Asia (India), Central and South America, Northern Australia and the tropical island of the Pacific and Indian oceans. The tropical seasonal forests are characterised by seasonal variations. The temperature and rains both are quite high, but the rainfall is not uniform throughout the year. The periodic rains (monsoons) result from the outflow of dry and hot winds from high pressure areas and an inflow of moisture-laden winds from the surrounding oceans.

The seasonal or deciduous forests are also known as monsoon forests. These are less lofty than the evergreen tropical forests and are almost leafless during dry season. The trees are tall well over 40 meters in height, but more commonly these are about 20-30 meters high. These are arranged in two strata. In the upper strata, the canopy is continuous and deciduous, but some trees in the lower strata are evergreen. These forests are rich in woody lianas and herbaceous epiphytes. Teak is the common tree and bamboo is the most common shrub.

The animal fauna is similar to tropical rain forests except that it is slightly less varied.

Subtropical Rainforests

The subtropical region of Florida in America has fairly high rain and is characterised by broad-leaved evergreen vegetation. The vegetation includes mahogany, gumbo, limbo bays, palms, oaks, mangolias and tamarind etc. All these are associated with epiphytes, ferns and vines. Animal life of subtropical forests is similar to rainforests.

***Animal Adjustments*:** The forest animals are comparatively strong and muscular and most of them are adapted to arboreal habit (for details see arboreal adaptations). These have a better sense of hearing and exhibit protective colouration. Their breeding habits are arboreal.

Grassland as a Part of Biomes

The grassland biomes are found both in the middle lattitude and equatorial tropical climatic zones. The major controlling factor is the rather sparse erratic rainfall of about 10-40 inches annually. The grasses vary from short buffallo grass to tall elephant grass and bamboo. There are various regional designations for the grassland biomes:

(1) *The Savannas:* These are found in the tropical plains of Africa, India and South America. The climate is characterised by a long dry winter season alternating with a wet summer season. The temperature remains relatively warm throughout the year.

(2) *The Prairies:* These are widespread in Central U.S.A., Eastern Europe and in the South American countries of Argentina and Uruguay. There are great annual variations in temperature because of their middle lattitude location, and rainfall tends to be scanty. The grasses vary from tall to short species.

(3) *The Steppe:* The steppe is drier than the prairie, contains short grass species and is located in interior plains and plateaues that are semiarid. Steppe grasslands are found in Western North America, but are most extensive in Central and West Asia.

***Fauna*:** The inhabitants of grassland are antelopes, hippopotamus, horses, asses, dogs, wild beasts, ground squirrels, cattle which graze upon grass, frogs, limbless amphibians, reptiles including turtles, tortoises, lizards and snakes and the grain eating birds. Insects, grass-hoppers, termites, ants, bees and wasps are dominant varieties.

The inhabitants of grassland are characterised by the power of swift running, flocking habit and keen sense of sight.

Desert as a Part of Biomes

Desert is that arid region of the earth surface, where rainfall and moisture are insufficient to support normal life. In these regions average rainfall never exceeds 10-15 inches per year. The rains are uncertain and may not occur even for years together. Depending upon the rains, deserts can be categorised into:

(1) low rainfall deserts *e.g.* Atacama,

(2) cold deserts such as Alps and Scandinavian mountains,

(3) hot deserts such as found in equatorial region,

(4) low nutrient deserts such as North American deserts,

(5) deserts due to toxic conditions as found near volcanoes and

(6) high salt deserts found in Chile, South America and Australia.

Deserts of the World

Deserts occupy about one-seventh of the land surface of the earth and are largely confined to world wide belt between Tropic of Cancer and the Tropic of Capricon. Deserts are found in almost all the countries of the world. Some of them are widely expanded, whereas others have a short range.

1. The Great North American desert covers Western United States, North Mexico, Central Mexico, South East California, South West Texas, Arizona and Nevada.
2. South American deserts are found in Atacama and Argentina. The former extends along the coast of Peru and Northern Chile and the later includes semi-arid regions of Andes.
3. In Europe the deserts are restricted to Alps and Scandinavian mountains.
4. The Asian deserts are Thar desert in North-West India (Rajasthan), Indus Valley in Pakistan and Gobi desert in China.

5. The African deserts are Sahara in Northern part and Kalahari in Southern region.
6. In Australia the central and western parts of the continent have desert conditions.

Desert as a Part of Environment

Desert climate is characterised by following features:

1. *Scarcity of Water:* Scarcity of water or dryness is characteristic of deserts. This is due to rainfall and absence of natural sources of water.
2. *Extremes of Temperature:* The temperature in day time is very high and relative humidity is very low. The scorching sunlight falls directly on living organisms which becomes unbearable. During night the temperature goes down tremendously because of the radiation of heat into space. Thus deserts present both the extremes of temperature within one day.
3. *Dust Storms:* Lack of moisture and streams of temperature cause an increase in the air movement and dust storms almost every evening.
4. *Lack of Vegetation:* Due to scarcity of water, plants are unable to grow. These occur only around areas, where very little water is available. The vegetation is also modified for this mode of life being succulent and thorny, which prevents excessive evaporation of water and helps in its conservation.

Plantation in Desert

The aridity of desert severely limits plant growth. The vegetation is very thin and is mainly found near oases. The plants which grow in desert are either drought-evading (ephemerals) or drought-resisting.

(i) The drought evading plants or ephemerals lie dormant as seeds and sprout only when moisture and temperature are favourable. Such plants have brief life-cycle to be completed within weeks. The African Boerhavia repens produces seeds even after eight or ten days of sprouting.

(ii) The drought resisting plants have evolved several means of storing water, locating underground water or reducing evaporation of water. Typical drought resisting plants are succulents which have very small or even no leaves. The steam is succulent and stores water. Their surface is covered by thick waterproof skin and the leaf-pores can be tightly shut. Some of them store water in tuberous roots. Most common examples of desert succulents are cacti and euphorbias, whose fleshy tissues hold vast amounts of liquid, and whose shallow, widespread roots absorb surface moisture.

(iii) Another variety of drought resistors are the shrubs such as creosote bushes, acacias, cassias and sage bushes etc. They possess short basal trunk with branches having short and thick leaves and thorns.

Animals Found in Desert

Under the harsh and shimmering sun the arid baked sand of the desert may appear to be devoid of animal life, but surprisingly large number of animal species thrive there. Like plants, these are also either drought evaders or drought resistors. Like ephemeral plants, drought evader animals also make their appearance only when rain sets in, for the rest of the time they remain dormant either in egg or pupal stage. Insects like crickets, grasshoppers, ants, bees, wasps, butterflies, moths and beetles swarm in the desert, when rains arrive and plants flourish. Tiny shrimps appear in temporary ponds, and lizards and snakes are also very common. Birds are few and live in places which have surface water for drinking within their flying range. Small rodents such as kangaroo rat, pocket mice and desert mangoose, desert fox, desert hedgehog, kit-fox, Jace-rabbit and the camels, deers, antelopes and asses are the mammalian forms found in deserts.

Adaptation According Desert-places

The fauna and flora of desert is peculiarly adapted to these drastic conditions and the modifications are mostly associated with the conservation and availability of water, protection against extreme heat and cold and obtaining food.

1. *Obtaining Water:* Higher animals cannot live long without drinking water and need water at frequent intervals. But scarcite of water is the primary cause for desert conditions. Free water is not available for miles together. Therefore, desert animals try to utilise water from every available source.
   (i) In most deserts water holes or springs are present as in Australia. Surface dwelling animals make use of this water and conserve it for long. If this water is not available these animals are unable to survive.
   (ii) A few animals absorb dew drops along with plants food.
   (iii) Certain desert animals such as desert rabbit, turtles and wood rat (Neotoma) do not drink water even if it is available to them. These suffice their need by eating succulent plans.
   (iv) Still another group or desert animals like Jerbous, kangaroo rat (Dipodomys), pocket mouse (Perognathus) and other desert rodents can live on the water received from eating dry seeds and vegetation.
   (v) Carnivorus animals get needful water from the body of their prey. The grasshopper-mice of western North America, long eared hedgehog of Sahara and mulgara, a marsupial which inhabits the most arid zone eat insects and small rodents.
   (vi) The most specialised group of animals (ant and rats-Jerboas and kangaroo rat) depends entirely upon metabolic water produced in their body during the oxidation of foodstuffs.

The oxidation of fat releases maximum water while minimum amount is released in the oxidation of proteins.

1 *g* of fat yields 1.1 gram water,

1 *g* carbohydrate yields 0.6 gram water, and

1 *g* protein yields 0.3 gram water.

(vii) Skin of certain animals like Moloch is hygroscopic and absorbs water like a blotting paper.

(viii) The young ones of several rodents are born at a time that coincides with the short rainy season so that the fluid lost through milk production is compensated by juicy grass.

2. *Water Conservation:* Terrestrial animals may lose water from their body in three ways:

(a) By evaporation through the body surface;

(b) By exhaling moisture during respiration; and

(c) By expelling water or liquid urine during excretion. Animals living in desert have developed means to avoid water loss by all these ways.

(a) *To Avoid Evaporation Through Body Surface:* Desert animals have developed several protective measures to avoid loss of water by evaporating through general body surface:

(i) Desert animals develop thick hide which avoids water loss by perspiration.

(ii) Number of sweat glands in the skin of desert mammals is very much reduced or in some cases these are totally absent.

(iii) Scorpions and reptiles have developed almost impermeable outer covering. Scales and spines are developed on the exposed surfaces. Examples are sand lizard (Moloch) and horn toad (Phymosoma).

(iv) Some desert animals produce epiphragms and secretions to prevent loss of water. Desert insects are wax-proof.

(v) Certain animals like camel have developed more tolerance to heat. Their body temperature fluctuates more with the atmosphere and thus reduces the amount of moisture lost through sweating and panting. Camel can lose 30 per cent of its body weight without showing signs of distress and protein deficiency. Mostly animals loose water from the blood which becomes thick or dehydrated causing death. In camel water is lost from the tissues and not from the blood.

(b) *To Avoid Loss of Moisture During Respiration:* Many desert mammals cool the exhaled air in their nose, before it is expelled out through nostrils. As a result moisture condenses in the nose and is not lost as water vapours along the exhaled air.

(c) *To Avoid Loss of Water During Excretion:* This is brought about by excreting concentrated urine or urine in semisolid state. Desert rats and kangaroo rats excrete urine which contains 24% urea, whereas in man urine contains just 6% urea. Even reptiles, insects and birds pass their body wastes in such concentrated state as uric acid that little or no moisture is lost in the process. Uric acid can be crystallised in the body. In camel also the amount of urea in the urine increases and the amount of urine decreases with the reduction in the amount of water conserved in the body.

(d) Other methods of conserving water:

(i) Desert animals are mostly burrowing and nocturnal. They come out of their burrows either late in night or early in the morning, when percentage of moisture in air is almost equal to the moisture in their burrow. The burrow protects them against scorching sunlight and avoids excessive water loss due to evaporation. SCHMDIT and NIELSEN'S have found that relative humidity in the burrows was 30-50% whereas outside the burrow in the air it was 0-15%. Certain animals are found to plug the mouths of their burrows during day time.

(ii) Desert animals are active only far a brief period during those part of the year when water and food are available. For rest of the time they remain dormant lying hidden in some burrows or holes.

(iii) Uromastix is said to store water in large intestine and in camel water is stored in the rumen along with fermenting food in a pocket-like structure. But the recent investigation suggest that the liquid present in rumen is not actually water but its composition is similar to digestive juices and its amount is far less than present in the stomach of other animals.

3. *Protection Against Scorching Sun:* Desert animals have to protect themselves against excessive heat. The temperature of the air may boost up upto 45°C and at that time the sand is still at a higher temperature and may be at 57-58°C. For protection against scorching sun and burning sand, desert animals have developed protective armours and several defensive mechanisms:

(i) Desert animals have thick hide or have a protective armour of scales, spines or dermal scutes.

(ii) Desert animals have long legs, which no doubt help in jumping and swift running, but also lift the body above the ground and thus avoid direct contact of the body with burning sand, *e.g.* leaf hoppers, kangaroo rat, and Jerboa etc.

(iii) The simplest and most wide-spread method of escaping desert heat is by burrowing. The burrows are humid and cooler. The air in the burrow acts as insulation. The desert animals hide themselves in the burrows during day time and come out only late in evening at night or early in the morning when temperature is low and relative humidity is more in the air.

(iv) Many desert animals have physical adaptations to enable them to burrow and survive under the sand. Scorpions, some lizards, certain beetles and desert mongoose are the burrowing animals. Scorpions have enlarged claws while lizards have wedge-shaped heads which cleave the sand. Certain species of beetles become so flattened that they can burrow in the sand with side to side movement of the body. Some lizards can dive in the sand with head first as in water.

(v) Some desert animals are acclimatised to the high temperature of desert and to restore their body temperature they bask in the sun. Beetles and lizards in general could endure temperature so high which cannot be tolerated even by human beings.

(vi) Other animals do not burrow but hide themselves under some suitable shelter and avoid the scorching sunrays of the noon.

4. *Protection Against Sand Storms:* The desert are often found to have dust storms in the afternoon and these are sometimes so strong that heaps of sand are washed away very swiftly. The important organs affected by the drifting sand are the nostrils, ears and eyes. Several mechanisms have been evolved to protect these structures:

(i) Nostrils in majority of desert animals are either reduced to small pinholes or they are protected with complicated valves. For example, in camel, the nostrils can be closed in the same way as eyes are closed by the eyelids. In sand-swimming lizards the nostrils are turned upwards and are thus protected from sand.

(ii) Eyelids exhibit various modifications and are capable of closing the eyes without hindering the vision. In Mabuya the lower eyelid is much enlarged and transparent and can be drawn on the eye like the nictitating membrane. It spreads over the complete eye thus protecting it against the sand but permitting vision. In certain other lizards (Ablepharus) the lower eyelid is permanently stretched over the eye and fused with the lower margin of upper eyelid. It is wholly transparent. Thus their eyes are permanently closed but are able to perceive vision. In camel the eyes are high on the head and are protected by long and thick eye lashes.

(iii) Ears are often protected with fringes of scales of hairs. In Phrynocephalus and Phrynosoma the air apertures are absent. Nocturnal animals need a sharp sense of hearing. Jack rabbits, hares etc. have large ears.

5. *Swift Running or Speed:* Since desert animals have to move far and wide in search of food and water, majority of them have great speed. In addition, for protection against enemies also these have to rely on speed. The limbs are specialised and adapted to walk on sand.

(i) The limbs are often long in desert animals, for example, leaf hopper and kangaroo rat. The latter has owned its name only on account of specially long hindlimbs.

(ii) Desert cat has wide soles thickly covered with fur which enables it to walk comfortably even on hot sand.

(iii) Legs of ostrich are padded with heavy callosities.

(iv) The undersurface of camel's feet is also padded.

(v) In Phrynocephalus and Teratonrhycus the digits possess lateral fringes.

(vi) In Scapteria the digits are broadened out into shovels.

(vii) American collar lizard (Crotaphytus collaris) and gridiron tailed lizard (Callisaurus) have longer hindlimbs and are capable of, running very fast. The latter is known to move on hindlimbs only with tail acting as a balancing organ.

6. *Colour:* The colouration in desert animals is found to match with the sand dunes. The body hues are often grey, brown or red, which is often in harmony with the ground. Often the colour is dull which blends with the surroundings. This blending colouration furnishes protection from the attack of predators by camouflaging. In the absence of definite places for concealment the colour-blending is the only measure of protection and the animals cannot be easily detected in moonlit nights. Warning colouration is exhibited by poisonous animals like lizard, Heloderma, rattle snake, spiders and red ants. In general, desert animals are less heavily pigmented and are comparatively small in size (Gloger's rule).

7. *Spinescence:* The spine studded body is characteristic feature of desert animals. The most conspicuous examples are Moloch horridus (flat spiny lizard), Phrynocephalus (horn toad) and Phrynosoma (horn lizard).

8. *Venom:* Possession of venom is another attribute to desert animals for self protection. Helodenna punctatum is the only poisonous lizard and is found in the arid zone of Mexico. Rattle snakes, trap door spider and Tarantula spiders are poisonous creatures of the desert.

9. *Sense Organs and Intelligentia:* Organs of sight, smell and hearing are specially developed. The jack rabbits and Jerboas have long external ears which collect and concentrate the sound waves from the air. These are able to see in the darkness of night. In desert mammals the tympanic bulla is enlarged. This probably adds greatly to the sensitivity of the animal's hearing. Certain desert animals exhibit very intelligent behaviour. Puff lizard when attacked, darts into a crevice between two rocks and inflates the body to such an extent, that it cannot be pulled out. Kangaroo rat when attacked by a rattle, throws sand and gravel in the face of the snake with the help of its hind-legs.

Reptiles and some insects appear to be preadapted to desert ecotype. Many reptiles have structural and behavioural adaptations to cope with sandy habit. Mammals as a group (except a few) are poorly adapted to desert life. But as such the desert fauna exhibits a very good example of adaptation to the environment.

Mountains as a Part of Biome

Mountains often compress many different biomes of life zones into a small area because of the differences in altitude. In Southern Mexico we can find almost all the ecosystems or biomes. At the lower altitudes, there are tropical rain forests. As one ascends, it is noted that at the eastern slope of the mountains the rain forest blends into the typical deciduous forest with a climate similar to that of deciduous biome. Continuing up the mountains is the coniferous forest with a climate similar to that of Southern Canada (coniferous forest zone). Finally is the area of perpetual ice and snow at the peaks of high mountains, which is comparable to polar regions of Tundra. The western slopes of mountains have desert condition at lower altitudes.

Animal life accords with the climatic conditions. Even the plant fauna is also similar. But there is some difference in the dispersion of animals. Except birds which have ability to fly, the other vertebrates do not reach the top of the mountains.

The land biomes are not always marked out sharply, but they exhibit overlappings and merging and they do not follow the lines of latitudes because of variation of altitude and variations in climate.

Caves as a Part of Biome

Caves are natural hollows under the earth surface or in mountains or hills. These actually represent abandoned channels through which underground rivers must have flowed in the past. Majority of caves are formed by the dissolution of limestone and are found only in those regions where massive limestones are deposited near the earth surface. Since the limestone is soluble in water, the surface water gradually sinks underground forming crevices, which increase in size sideways and form the channels, known as caves or caverns.

Caves of the World: Caves are found all over the world but are restricted to particular areas. The most famous cave area is located in the calcareous Alps of central and southern Europe extending through France to Austria and Yugoslavia. In America it extends through parts of Indiana. Kentucky, Tennessee and Missori. Well known caves are present in Himalayas, Malaysia, Indonesia, South Africa and Australia. Dry caves are present in the rocks of Gibralter, Malta and Lebanon. Caves are also found in Europe especially in Carniola, Austria and Illyria. Basses, Pyreneest in France and Spain, and Dordogne region in Western France and others are well known caves in which Palaeolithic man had lived.

Environmental Features of Caves

The cave environment presents stable conditions for living beings. The environmental conditions are more or less similar in different caves.

Caves are characterised by the absence of bright light and

nearly uniform temperature, high humidity, feeble fluctuations of temperature and absence of air currents. But light is present at the mouth of the caves which penetrates for a short distance inside. Therefore, the cave environment is separated by LULL into three distinct regions, namely:

(i) Twilight or dysphotic or transitional region.

(ii) Region of fluctuating temperature.

(iii) The inner cave region.

The latter represents true cavern environment having no light, relatively high humidity and absence of air currents. The absence of light has a pronounced effect on the inhabitants of this region.

Cave Flora: The absence of light does not favour life and growth of green plants which are used as food. Therefore, only those plants exist which can bear the darkness. These are Ozonium (common fungus), green mould, long white mould, mushrooms (Agaricus sp.).

Cave Fauna: Cave dwellers are classified into three groups:

1. *Permanent Cave Dwellers:* The permanent cave dwellers reside permanently in caves and do not come out of them. These include both terrestrial as well as aquatic forms and are represented from almost all the phyla: turbellarians from Platyhelminthes; chaetopod worms and leeches from Annelida; amphipods, isopods, decapods etc. from class Crustacea; springtails from class Insecta; mites and true spiders from Arachnida and snails from Gastropoda. The cave dweller vertebrates are represented by all the classes. Amongst fishes are Amblyopsis. Syphlichthyes chologaster and the cave catfish (Gronias nigrilabris). In addition, large number of fishes of the family Amblyopsidae are found in North American caves. Cave amphibians are cave salamanders such as Spelerpes maculicauda, Typhlotriton spelaeus, Proteus anguinus. Nocturnal owls and oil birds nest in the caves but are not

permanently cave dwellers. Amongst mammals white footed mouse (Peromyscus leucopus) is the only cave dwelling form.

2. *Occasional Cave-dwellers:* These constitute the temporary fauna of caves and include bats and owls.
3. Accidental cave dwellers or strugglers.

Cave Adaptations: The body organisation of permanent cave dwellers has undergone marked changes, but these are mostly towards degeneration. The obvious reason is that cave environment presents simplified ecological conditions, characterised by uniform and rather unchangeable atmosphere, lack of competition, absence of natural enemies. Therefore, cave animals are known for their primitive organisation and defenceless nature.

1. *Loss of Pigmentation:* Cave animals are either colourless or pale white in colour. The loss of pigmentation is on account of absence of light in the caves. If coloured, they are lighter or paler than their epigean relatives. Moreover, certain cave dwellers when exposed to light, develop pigments. For example, Niphargus (a cave dweller amphipod) when reared in light, develops pigment in a short period. Similarly, Proteus develops dark pigment when exposed to light. In certain cases the light is detrimental. Spring tails and mites may die when exposed to light.
2. *Slender Body and Reduced Appendages:* The cave dwelling forms are often found to possess elongated and slender body with attenuated appendages. For example, in crustaceans, Bathynella and Parabathynella, the antennules and antennae are slender. In urodeles the body is slender and limbs are short and slender.
3. *Weak Exoskeleton:* The cave arthropods have weak or fragile exoskeleton. The shell is thin in cave snail. In cave fishes scales are lost altogether. The insects dwelling in caves are thinner than their surface dwelling relatives.

4. *Absence of Visual Organs:* Since light is absent in caves the eyes are either reduced, degenerated or absent in cave animals, or these are embedded in the integument. The eyes are completely absent in cave fishes *e.g.* Stygicola and Leucifuga from Cuba and Amblyopsis from mammoth caves of America. Among amphibians eyes are nonfunctional in Typhlomolge and Typhlotriton and in Apodon and Ichthyophis these are covered by a fold of integument and appear as mere black dots.
5. *Tactile Organs:* To compensate the impaired sense of sight, other senses are highly developed in cave animals. The cave dwellers often depend either on smell or on touch. In cave insects, myriapods and arachnids the legs and antennae are long and provided with tactile hairs. The head, snout and lips of cave fish Amblyopsis bear taste organs in the form of terminal buds. Chemoreceptors are well developed in cave animals than in their epigean relatives. For example, cave isopod, Asellus has large sense bulbs.
6. *Retrogression of Endocrine System:* In cave animals the endocrine system exhibits varying degree of degeneration, especially in cave urodeles. The thyroid gland in Proteus is said to secrete a hormone which has no effect on the body tissues.

In addition to the above modifications the cave animals are characterised by low metabolic rate, very low fecundity and slow rate of development. There is no competition or struggle for existence among cave dwellers. Hence instead of specialisation, these have under gone degeneration. Further the environment is roughly constant with no major changes, even for hundreds of years together. These are lethargic creatures.

Therefore, archaic forms have been able to establish themselves. The snails of the genus Lartetia, Bathynella, Proteus (salamander), Troglochaetus (annelid), and Cirolanides (an isopod) are reptiles of the past. The crustacean Bathynella belongs to palaeozoic group, Syncardia. Cave dwellers possess such primitive features as presence of lateral line system on the head of Caecobarbus and the ossification of hyoid and branchial arches in Proteus (an amphibian). Neotenic forms are also found in cave dwellers *e.g.* Proteus and Siren.

Cause of Diversity

It is the basic source of biodiversity. The genes found in organisms can form enormous number of combinations each of which gives rise to some variability. Genes are the basic units of hereditary information transmitted from one generation to other. When the genes within the same species show different versions due to new combinations, it is called genetic variability. For example, all rice varieties belong to the species *Oryza sativa,* but there are thousands of wild and cultivated varieties of rice which show variations at the genetic level and differ in their colour, size, shape, aroma and nutrient content of the grain. This is the genetic diversity of rice.

Classification of Species

This is the variability found within the population of a species or between different species of a community. It represents broadly the species richness and their abundance in a community. There are two popular indices of measuring species diversity known as Shannon- Wiener index and Simpson index. What is the number of species on this biosphere?

The estimates of actual number vary widely due to incomplete and indirect data. The current estimates given by Wilson in 1992 put the total number of living species in a range of 10 million to 50 million. Till now only about 1.5 million living and 300,000 fossil species have been actually described and given scientific names. It is quite likely that a large fraction of these species may become extinct even before they are discovered and enlisted.

Diversity in Ecosystem: This is the diversity of ecological complexity showing variations in ecological niches, trophic structure, food-webs, nutrient cycling, etc. The ecosystems also show variations with respect to physical parameters like moisture, temperature, altitude, precipitation, etc. Thus, there occurs tremendous diversity within the ecosystems, along these gradients. We may consider diversity in forest ecosystem, which is supposed to have mainly a dominance of trees. But, while considering a tropical rainforest, a tropical deciduous forest, a temperate deciduous forest and a boreal forest, the variations observed are just too many and they are mainly due to variations in the above-mentioned physical factors.

The ecosystem diversity is of great value that must be kept intact. This diversity has developed over millions of years of evolution. If we destroy this diversity, it would disrupt the ecological balance. We cannot even replace the diversity of one ecosystem by that of another. Coniferous trees of boreal forests cannot take up the function of the trees of tropical deciduous forest lands and vice versa, because ecosystem diversity has evolved with respect to the prevailing environmental conditions with well-regulated ecological balance.

Classification of India on Biogeographical Grounds: India has different types of climate and topography in different parts of the country and these variations have induced enormous variability in flora and fauna. India has a rich heritage of biological diversity and occupies the tenth position among the plant rich nations of the world.

It is very important to study the distribution, evolution, dispersal and environmental relationship of plants and animals in time and space. Biogeography comprising of phytogeography and zoogeography deals with these aspects of plants and animals. In order to gain insight about the distribution and environmental interactions of flora and fauna of our country, it has been classified into ten biogeographic zones as given below. Each of these zones has its own characteristic climate, soil, topography and biodiversity.

India's Major Biogeographic Habitats

Sr. No.	Biogeographic Zone	Biotic Province	Total area (Sq. Km.)
1.	Trans-Himalayan	Upper Regions	186200
2.	Himalayan	North-West Himalayas	6900
		West Himalayas	720000
		Central Himalayas	123000
		East Himalayas	83000
3.	Desert	Kutch	45000
		Thar	180000
		Laddakh	NA
4.	Semi-Arid	Central India	107600
		Gujarat-Rajwara	400400
5.	Western Ghats	Malabar Coast	59700
		Western Ghat Mountains	99300
6.	Deccan Peninsula	Deccan Plateau South	378000
		Central Plateau	341000
		Eastern Plateau	198000
		Chhota Nagpur	217000
		Central Highlands	287000
7.	Gangetic Plain	Upper Gangetic Plain	206400
		Lower Gangetic Plain	153000
8.	North-East India	Brahmaputra Valley	65200
		North-Eastern Hills	106200
9.	Islands	Andaman Islands	6397
		Nicobar Islands	1930
		Lakshadweep Islands	180
10.	Coasts	West Coast	6500
		East Coast	6500

Source: "Conserving our Biological Wealth", WWF for Nature-India and Zoological Survey of India.

Biodiversity's Value: The value of biodiversity in terms of its commercial utility, ecological services, social and aesthetic value is enormous. We get benefits from other organisms in innumerable ways. Sometimes we realise and appreciate the value of the organism only after it is lost from this earth. Very small, insignificant, useless looking organism may play a crucial role in the ecological balance of the ecosystem or may be a potential source of some invaluable drug for dreaded diseases like cancer or AIDS. The multiple uses of biodiversity or biodiversity value has been classified by McNeely *et al* in 1990 as follows:

(i) *Consumptive use Value:* These are direct use values where the biodiversity product can be harvested and consumed directly *e.g.* fuel, food, drugs, fibre, etc.

Food: A large number of wild plants are consumed by human beings as food. About 80,000 edible plant species have been reported from wild. About 90% of present-day food crops have been domesticated from wild tropical plants. Even now our agricultural scientists make use of the existing wild species of plants that are closely related to our crop plants for developing new hardy strains. Wild relatives usually possess better tolerance and hardiness. A large number of wild animals are also our sources of food.

Drugs and Medicines: About 75% of the world's population depends upon plants or plant extracts for medicines. The wonder drug Penicillin used as an antibiotic is derived from a fungus called Penicillium. Likewise, we get Tetracycline from a bacterium. Quinine, the cure for malaria is obtained from the bark of Cinchona tree, while Digitalin is obtained from foxglove (Digitalis) which is an effective cure for heart ailments. Recently vinblastin and vincristine, two anticancer drugs, have been obtained from Periwinkle (Catharanthus) plant, which possesses anticancer alkaloids. A large number of marine animals are supposed to possess anti-cancer properties which are yet to be explored systematically.

Fuel: Our forests have been used since ages for fuel wood. The fossil fuels coal, petroleum and natural gas are also products of fossilised biodiversity. Firewood collected by individuals are not normally marketed, but are directly consumed by tribals and local villagers, hence falls under consumptive value.

(ii) *Productive use Values:* These are the commercially usable values where the product is marketed and sold. It may include lumber or wild gene resources that can be traded for use by scientists for introducing desirable traits in the crops and domesticated animals. These may include the animal products like tusks of elephants, musk from musk deer, silk from silk-worm, wool from sheep, fir of many animals, lac from lac insects, etc. all of which are traded in the market. Many industries are dependent upon the productive use values of biodiversity *e.g.*- the paper and pulp industry, plywood industry, railway sleeper industry, silk industry, textile industry, ivory-works, leather industry, pearl industry, etc.

Despite international ban on trade in products from endangered species, smuggling of fur, hide, horns, tusks, live specimen, etc. worth millions of dollars are being sold every year. Developing countries in Asia, Africa and Latin America are the richest biodiversity centers and wildlife products are smuggled and marketed in large quantities to some rich Western countries and also to China and Hong Kong where export of cat skins and snake skins fetches a booming business.

(iii) *Social Value:* These are the values associated with the social life, customs, religion and psycho-spiritual aspects of the people. Many of the plants are considered holy and sacred in our country like Tulsi (holy basil), Peepal, Mango, Lotus, Bael, etc. The leaves, fruits or flowers of these plants are used in worship or the plant itself is worshipped. The tribal people are very closely linked with the wildlife in the forests. Their social life, songs, dances and customs are closely woven around the wildlife. Many animals like cow, snake, bull, peacock, owl, etc. also have significant place in our psycho-spiritual arena and thus hold special social importance. Thus biodiversity has distinct social value, attached with different societies.

(iv) *Ethical Value:* It is also sometimes known as existence value. It involves ethical issues like "all life must be preserved". It is based on the concept of "Live and Let Live". If we want our human race to survive, then we must protect all biodiversity, because biodiversity is valuable.

The ethical value means that we may or may not use a species, but knowing the very fact that this species exists in nature gives us pleasure. We all feel sorry when we learn that "passenger pigeon" or "dodo" is no more on this earth. We are not deriving anything direct from kangaroo, zebra or giraffe, but we all strongly feel that these species should exist in nature. This means, there is an ethical value or existence value attached to each species.

(v) *Aesthetic Value:* Great aesthetic value is attached to biodiversity. No one of us would like to visit vast stretches of barren lands with no signs of visible life. People from far and wide spend a lot of time and money to visit wilderness areas where they can enjoy the aesthetic value of biodiversity and this type of tourism is now known as eco-tourism. The "Willingness to pay" concept on such eco-tourism gives us even a monetary estimate for aesthetic value of biodiversity. Eco-tourism is estimated to generate about 12 billion dollars of revenue annually, that roughly gives the aesthetic value of biodiversity.

(vi) *Option Values:* These values include the potentials of biodiversity that are presently unknown and need to be explored. There is a possibility that we may have some potential cure for AIDS or cancer existing within the depths of a marine ecosystem, or a tropical rainforest.

Thus option value is the value of knowing that there are biological resources existing on this biosphere that may one day prove to be an effective option for something important in the future. Thus, the option value of biodiversity suggests that any species may prove to be a miracle species someday. The biodiversity is like precious gifts of nature presented to us. We should not commit the folly of losing these gifts even before unwrapping them.

The option value also includes the values, in terms of the option to visit areas where a variety of flora and fauna, or specifically some endemic, rare or endangered species exist.

(vii) *Ecosystem Service Value:* Recently, a non-consumptive use value related to self-maintenance of the ecosystem and various important ecosystem services has been recognised. It refers to the services provided by ecosystems like prevention of soil erosion, prevention of floods, maintenance of soil fertility, cycling of nutrients, fixation of nitrogen, cycling of water, their role as carbon sinks, pollutant absorption and reduction of the threat of global warming, etc.

Different categories of biodiversity value clearly indicate that ecosystem, species and genetic diversity all have enormous potential and a decline in biodiversity will lead to huge economic, ecological and socio-cultural losses.

Biodiversity Value of Some Selected Organisms in Monetary Terms: A male lion living upto an age of 7 years can generate upto $ 515,000 due to its aesthetic value as paid by tourists, whereas if killed for the lion skin a market price upto $ 1,000 can be fetched.

In its lifetime a Kenyan elephant can earn worth $ 1 million as tourist revenue. The mountain gorillas in Rwanda are fetching $ 4 million annually through eco-tourism. Whale watching on Hervey Bay on Queensland's coast earns $12 million annually.

Tourism to Great Barrier Reef in Australia earns $ 2 billion each year. A typical tree provides $ 196,2150 worth of ecological services as oxygen, clean air, fertile soil, erosion control, water recycling, wildlife habitat, toxic gas moderation, etc. Whereas its worth is only about $ 590 if sold in the market as timber.

Biodiversity at International Level: Following the 1992 "Earth Summit" at Rio de Janeiro, it became evident that there is a growing need to know and scientifically name, the huge number of species which are still unknown on this earth. Roughly 1.5 million species are known till date which is perhaps 15% or may be just 2% of the actual number. Tropical deforestation alone is reducing the biodiversity by half a per cent every year. Mapping the biodiversity has therefore, been rightly recognised as an emergency task in order to plan its conservation and practical utilisation in a judicious manner.

Terrestrial biodiversity of the earth is best described as biomes, which are the largest ecological units present in different geographic areas and are named after the dominant vegetation *e.g.* the tropical rainforests, tall grass prairies, savannas, desert, tundra, etc. The tropical rainforests are inhabited by teeming millions of species of plants, birds, amphibians, insects as well as mammals.

They are the earth's largest storehouse of biodiversity. Many of these species have developed over the time in highly specialised niches and that makes them more vulnerable to extinction when their natural home or niche is destroyed. About 50 to 80% of global biodiversity lies in these rainforests.

More than one-fourth of the world's prescription drugs are extracted from plants growing in tropical forests. Out of the 3000 plants identified by National Cancer Research Institute as sources of cancer-fighting chemicals, 70% come from tropical rain forests.

Very recently, extract from one of the creeping vines in the rainforests at Cameroon has proved effective in the inhibition of replication of AIDS virus. It is interesting to note that the common neem tree, so popular in tropical India, known for its medicinal properties has now come into lime light even in the Western temperate countries.

There is an estimated 1,25,000 flowering plant species in tropical forests. However, till now we know only 1-3% of these species. Needless to say, we must try in every way to protect our tropical rainforests.

The Silent Valley in Kerala is the only place in India where tropical rain forests occur. You may recall the case of Silent Valley Hydroelectric Project, which was abandoned mainly because it had put to risk our only tropical rain forest biodiversity.

Temperate forests have much less biodiversity, but there is much better documentation of the species. Globally, we have roughly 1,70,000 flowering plants, 30,000 vertebrates and about 2,50,000 other groups of species that have been described. There is a stupendous task of describing the remaining species which may range anywhere from 8 million to 100 million.

The below-given table shows the estimated number of some known living species in different taxonomic groups:

Living Species Estimates (World Resource Institute, 1999)

Taxonomic group	*Number*
Bacteria & Cyanobacteria	5,000
Protozoans (Single-celled animals)	31,000
Algae	27,000
Fungi (Molds, Mushrooms)	45,000
Higher Plants	2,50,000
Sponges	5,000
Jelly fish, Corals, etc.	10,000
Flatworms, roundworms, earthworms	36,000
Snails, Clams, Slugs, etc.	70,000
Insects	7,50,000
Mites, Ticks, Croaks, shrimps	1,20,000

Contd....

Taxonomic group	*Number*
Fish and Sharks	22,000
Amphibians	4,000
Reptiles	5,000
Birds	9,000
Mammals	4,000
Total	1,400,000

It is interesting to know that marine diversity is even much higher than terrestrial biodiversity and ironically, they are still less known and described. Estuaries, coastal waters and oceans are biologically diverse and the diversity is just dazzling. Sea is the cradle of every known animal phylum. Out of the 35 existing phyla of multicellular animals, 34 are marine and 16 of these are exclusively marine.

National Level Bio-diversity (Indian Biodiversity): Every country is characterised by its own biodiversity depending mainly on its climate. India has a rich biological diversity of flora and fauna. Overall six per cent of the global species are found in India. It is estimated that India ranks 10th among the plant rich countries of the world, 11th in terms of number of endemic species of higher vertebrates and 6th among the centers of diversity and origin of agricultural crops. The total number of living species identified in our country is 150,000. Out of a total of 25 biodiversity hot-spots in the world, India possesses two, one in the north-east region and one in the western ghats. India is also one of the 12 mega-biodiversity countries in the world, which will be discussed later.

Biodiversity at Local or Regional Levels: Biodiversity at regional level is better understood by categorizing species richness into four types, based upon their spatial distribution as discussed below:

(i) Point richness refers to the number of species that can be found at a single point in a given space.

(ii) Alpha (α-) richness refers to the number of species found in a small homogeneous area.

(iii) Beta (β-) richness refers to the rate of change in species composition across different habitats.

(iv) Gamma (γ-) richness refers to the rate of change across large landscape gradients.

α-richness is strongly correlated with physical environmental variables.

For example, there are 100 species of tunicates in arctic waters, 400 species in temperate waters and 600 in tropical seas.

Thus, temperature seems to be the most important factor affecting α-richness of tunicates. β-richness means that the cumulative number of species increases as more heterogeneous habitats are taken into consideration.

For example, the ant species found in local regions of north pole is merely 10. As we keep on moving towards the equator and thus add more and more habitats, the number of species of ants reaches as high as 2000 on the equatorial region.

India's Status as a Mega Diversity Nation: India is one of the 12 mega-diversity countries in the world. The Ministry of Environment and Forests, Govt. of India (2000) records 47,000 species of plants and 81,000 species of animals which is about 7% and 6.5% respectively of global flora and fauna. Distribution of species in some major groups of flora and fauna in India:

Group-wise Species Distribution

Plants	*Number*	*Animals*	*Number*
Bacteria	850	Lower groups	9979
Fungi	23,000	Mollusca	5042
Algae	2500	Arthropoda	57,525
Bryophytes	2564	Pisces (Fishes)	2546
		Amphibia	

Contd...

Plants	*Number*	*Animals*	*Number*
Pteridophytes	1022	Reptiles	428
Gymnosperms	64	Birds	1228
Angiosperms	15,000		204
		Mammals	372

Endemism: Species which are restricted only to a particular area are known as endemic. India shows a good number of endemic species. About 62% of amphibians and 50% of lizards are endemic to India. Western ghats are the site of maximum endemism.

Centre of Origin: A large number of species are known to have originated in India. Nearly 5000 species of flowering plants had their origin in India. From agro-diversity point of view also our country is quite rich.

India has been the centre of origin of 166 species of crop plants and 320 species of wild relatives of cultivated crops, thereby providing a broad spectrum of diversity of traits for our crop plants.

Marine Diversity: Along 7500 km long coastline of our country in the mangroves, estuaries, coral reefs, back waters, etc. there exists a rich biodiversity.

More than 340 species of corals of the world are found here. The marine diversity is rich in mollusks, crustaceans (crabs, etc.), polychaetes and corals. Several species of mangrove plants and sea-grasses (Marine algae) are also found in our country. A large proportion of the Indian Biodiversity is still unexplored.

There are about 93 major wet lands, coral reefs and mangroves which need to be studied in detail. Indian forests cover 64.01 million hectares having a rich biodiversity of plants in the Trans-Himalayan, north-west, west, central and eastern Himalayan forests, western ghats, coasts, deserts, Gangetic plains, Deccan plateau and the Andaman, Nicobar and

Lakshadweep islands. Due to very diverse climatic conditions there is a complete rainbow spectrum of biodiversity in our country.

Biodiversity—Main Places: *Areas which exhibit high species richness as well as high species endemism are termed as hot spots of biodiversity.* The term was introduced by Myers (1988). There are 25 such hot spots of biodiversity on a global level out of which two are present in India, namely the Eastern Himalayas and Western Ghats. These hotspots covering less than 2% of the world's land area are found to have about 50% of the terrestrial biodiversity. According to Myers *et al.* (2000) an area is designated as a hotspot when it contains at least 0.5% of the plant species as endemics.About 40% of terrestrial plants and 25% of vertebrate species are endemic and found in these hotspots. After the tropical rain forests, the second highest number of endemic plant species are found in the Mediterranean (Mittermeier).

Broadly, these hot spots are in Western Amazon, Madagascar, North and East Borneo, North Eastern Australia, West Africa and Brazilian Atlantic forests. These are the areas of high diversity, endemism and are also threatened by human activities.

More than 1 billion people (about 1/6th of the world's population), most of whom are desperately poor people, live in these areas.

Any measures of protecting these hotspots need to be planned keeping in view the human settlements and tribal issues. Earlier 12 hot spots were identified on a global level. Later Myers *et al* (2000) recognised 25 hot spots.

Two of these hotspots lie in India extending into neighbouring countries namely, Indo-Burma region (covering Eastern Himalayas) and Western Ghats- Sri Lanka region. The Indian hotspots are not only rich in floral wealth and endemic species of plants but also reptiles, amphibians, swallow tailed butterflies and some mammals.

Global Hotspots of Biodiversity

Hotspots	*Plant Species*	*Endemic Plants*	*%of Global Plants*	*Vertebrate Species*	*Endemic Vertebrates*	*%of Global Vertebrates*
1. Tropical Andes	45000	20000	6.7	3389	1567	5.7
2. Mesoamerican forests	24000	5000	1.7	2859	1159	4.2
3. Caribbean	12000	7000	2.3	1518	779	2.9
4. Brazil's Atlantic Forest	20000	8000	2.7	1361	567	2.1
5. Choc/Darien of Panama Western Ecuador	9000	2250	0.8	1625	418	1.5
6. Brazil's Cerrado	10000	4400	1.5	1268	117	0.4
7. Central Chile	3429	1605	0.5	335	61	0.2
8. California Flonstic Province	4426	2125	0.7	584	71	0.3
9. Madagascar	12000	9704	3.2	987	771	2.8
10. Eastern Arc and Coastal Forest of Tanzania/Kenya	4000	1500	0.5	1019	121	0.4
11. Western African Forests	9000	2250	0.8	1320	270	1.0
12. Cape Floristic Province	8200	5682	1.9	562	53	0.2

Contd...

Hotspots	***Plant Species***	***Endemic Plants***	***%of Global Plants***	***Vertebrate Species***	***Endemic Vertebrates***	***%of Global Vertebrates***
13. Succulent Karoo	4849	1940	0.6	472	45	0.2
14. Mediterranean Basin	25000	13000	4.3	770	235	0.9
15. Caucasus	6300	1600	0.5	632	59	0.2
16. Sundaland	25000	15000	5.0	1800	701	2.6
17. Wallacea	10000	1500	0.5	1142	529	1.9
18. Philippines	7620	5832	1.9	1093	518	1.9
19. Indo-Burma						
Eastern Himalayas	13500	7000	2.3	2185	528	1.9
20. South-Central China	12000	3500	1.2	1141	178	0.7
21. Western-Ghats Sri Lanka	4780	2180	0.7	1073	355	13
22. South-western Australia	5469	4331	1.4	456	100	0.4
23. New Caledonia	3332	2551	0.9	190	84	0.3
24. New Zealand	2300	1865	0.6	217	136	0.5
25. Polynesia/Micronesia	6557	3334	1.1	342	223	0.8
Total	—	133,149	44.4	—	9645	35.3

Source: Myers *et al*, 2000.

(a) *Eastern Himalayas:* They display an ultra-varied topography that fosters species diversity and endemism. There are numerous deep and semi-isolated valleys in Sikkim which are extremely rich in endemic plant species. In an area of 7298 Km^2 of Sikkim about 4250 plant species are found of which 60% are endemic.

The forest cover of Eastern Himalayas has dwindled to about 1/3rd of its original cover. Certain species like *Sapria himalayana,* a parasitic angiosperm was sighted only twice in this region in the last 70 years.

Recent studies have shown that North East India along with its contiguous regions of Burma and Chinese provinces of Yunnan and Schezwan is an active centre of organic evolution and is considered to be the cradle of flowering plants. Out of the world's recorded flora 30% are endemic to India of which 35,000 are in the Himalayas.

(b) *Western Ghats:* It extends along a 17,000 Km^2 strip of forests in Maharashtra, Karnataka, Tamil Nadu and Kerala and has 40% of the total endemic plant species. Sixty-two per cent amphibians and 50% lizards are endemic to Western Ghats.

Forest tracts upto 500 m elevation covering 20% of the forest expanse are evergreen while those in 500-1500 m range are semi-evergreen. The major centers of diversity are *Agastyamalai Hills* and *Silent* Valley—the New Amambalam Reserve Basin.

It is reported that only 6.8% of the original forests are existing today while the rest has been deforested or degraded, which raises a serious cause of alarm, because it means we have already lost a huge proportion of the biodiversity.

Although the hotspots are characterised by endemism, interestingly, a few species are common to both the hotspots in India. Some common plants include *Ternstroemia japonica, Rhododendron* and *Hypericum,* while the common fauna includes laughing thrush, fairy blue bird, lizard hawk, etc. indicating their common origin long back in the geological times.

Challenges to Biodiversity: Extinction or elimination of a species is a natural process of evolution. In the geologic period the earth has experienced mass extinctions. During evolution, species have died out and have been replaced by others. However, the rate of loss of species in geologic past has been a slow process, keeping in view the vast span of time going back to 444 million years.

The process of extinction has become particularly fast in the recent years of human civilization. In this century, the human impact has been so severe that thousands of species and varieties are becoming extinct annually. *One of the estimates by the noted ecologist, E.O. Wilson puts the figure of extinction at 10,000 species per year or 27 per day!* This startling figure raises an alarm regarding the serious threat to biodiversity. Over the last 150 years the rate of extinction has escalated more dramatically. If the present trend continues we would lose 1/3rd to 2/3rd of our current biodiversity by the middle of twenty-first century. Let us consider some of the major causes and issues related to threats to biodiversity.

Habitats in Annihilation

Destruction and loss of natural habitat is the single largest cause of biodiversity loss. Billions of hectares of forests and grasslands have been cleared over the past 10,000 years for conversion into agriculture lands, pastures, settlement areas or development projects. These natural forests and grasslands were the natural homes of thousands of species which perished due to loss of their natural habitat. Severe damage has been caused to wetlands thinking them to be useless ecosystems. The unique rich biodiversity of the wetlands, estuaries and mangroves are under the most serious threat today.

The wetlands are destroyed due to draining, filling and pollution thereby causing huge biodiversity loss. Sometimes the loss of habitat is in installments so that the habitat is divided into small and scattered patches, a phenomenon known as habitat fragmentation. There are many wildlife species such as bears and large cats that require large territories to subsist. They get badly threatened as they breed only in the interiors of the forests. Due to habitat fragmentation many song birds are vanishing.

There has been a rapid disappearance of tropical forests in our country also, at a rate of about 0.6% per year. With the current rate of loss of forest habitat, it is estimated that 20-25% of the global flora would be lost within a few years. Marine biodiversity is also under serious threat due to large-scale destruction of the fragile breeding and feeding grounds of our oceanic fish and other species, as a result of human intervention.

Poaching: Illegal trade of wildlife products by killing prohibited endangered animals *i.e.* poaching is another threat to wildlife. Despite international ban on trade in products from endangered species, smuggling of wildlife items like furs, hides, horns, tusks, live specimens and herbal products worth millions of dollars per year continues. The developing nations in Asia, Latin America and Africa are the richest source of biodiversity and have enormous wealth of wildlife. The rich countries in Europe and North America and some affluent countries in Asia like Japan, Taiwan and Hong Kong are the major importers of the wildlife products or wildlife itself.

The trading of such wildlife products is highly profit making for the poachers who just hunt these prohibited wildlife and smuggle it to other countries mediated through a mafia. The cost of elephant tusks can go upto $ 100 per kg; the leopard fur coat is sold at $ 100,000 in Japan while bird catchers can fetch upto $ 10,000 for a rare hyacinth macaw, a beautiful coloured bird, from Brazil. The worse part of the story is that for every live animal that actually gets into the market, about 50 additional animals are caught and killed.

If you are fond of rare plants, fish or birds, please make sure that you are not going for the endangered species or the wild-caught species. Doing so will help in checking further decline of these species. Also do not purchase furcoat, purse or bag, or items made of crocodile skin or python skin. You will certainly help in preserving biodiversity by doing so.

Significance of Biodiversity

If we divide the whole earth's mass into 10 billion parts, it is only in one part where life exists and the astounding variety of living organisms numbering somewhere around 50 million species are all restricted to just about a kilometer-thick layer of soil, water and air. Isn't it wonderful to see that so much diversity has been created by nature on this earth from so little physical matter! Biodiversity refers to the variety and variability among all groups of living organisms and the ecosystem complexes in which they occur.

From the driest deserts to the dense tropical rainforests and from the high snow-clad mountain peaks to the deepest of ocean trenches, life occurs in a marvellous spectrum of forms, size, colour and shape, each with unique ecological inter-relationships. Just imagine how monotonous and dull the world would have been had there been only a few species of living organisms that could be counted on fingertips!

In the Convention of Biological diversity (1992) biodiversity has been defined as the variability among living organisms from all sources including inter alia, terrestrial, marine and other aquatic ecosystems and the ecological complexes of which they are a part.

Biodiversity and its Forms

Units of biodiversity may range from the genetic level within a species to the biota in a specific region and may extend up to the great diversity found in different biomes.

4

ELEMENTS AT WORK IN ECOLOGY

STRUCTURE OF ECOSYSTEM

Ecological pyramids represent the trophic structure and also the trophic function of an ecosystem. In an ecological pyramid, the first trophic level forms the base and successive trophic levels the tiers which make up the apex. Ecological pyramids may be of three general types, *i.e.*:(1) Pyramid of number, (2) Pyramid of biomass and (3) Pyramid of energy.

1. *Pyramid of Number:* It represents the numerical relationship between different trophic levels of a food chain. In such a pyramid, the more abundant species form the base of pyramid and the less abundant species remain near the top. The pyramid of number indicates that the producers (algae, diatoms or green plants) which form the base of the pyramid are ingested in large numbers by a comparatively small number of primary consumers or herbivores. These are eaten by a small number of primary carnivores or secondary consumers and these, in turn, are ingested by still small number of tertiary consumers or secondary

carnivores. The pyramid of numbers can be best understood by taking lake or grassland as an example.

In a lake ecosystem, the base of pyramid, *i.e.* the lowest trophic level is occupied by the producers such as diatoms and algae. These are more in abundance in comparison to the herbivores. The second trophic level is represented by zooplankton which are the primary consumers and are less abundant. The third trophic level is occupied by a still smaller number of carnivores, *i.e.* medium sized fishes and the apex by large sized carnivores, *i.e.* large sized fishes which are only a few in number. Similarly tiger in grassland ecosystem occupies the apex of the pyramid and the producers, *i.e.* green plants occupy the base. Thus in the pyramid of number there is a gradual loss in the number of organisms and an increase in the size of the body.

In food chain involving parasites, the pyramid of numbers is reversed for the successive steps of parasite dependency and the parasites are generally more numerous than their hosts. For example, a single tree supports a larger number of fruit eating herbivore birds which in turn support a still larger number of parasites like bugs and lice.

2. *Pyramid of Biomass:* The biomass, *i.e.* the living weight of the organisms of the food chain present at any time in an ecosystem forms the pyramid of biomass. The pyramid of biomass indicates the decrease or the gradual reduction in biomass at each trophic level from base to apex. In the terrestrial environment, the biomass of the producers, *i.e.* vegetation is usually the greatest.

   These occupy the basic trophic level in the food chain. The next trophic level is occupied by the primary consumers or the herbivores. The total biomass of producers, is much more than the total biomass of herbivores. Likewise the total biomass of the primary carnivores (secondary consumers) will be less than the herbivores and so on. Similarly, in marine or freshwater ecosystem, the basic trophic level is occupied by phytoplankton and zooplankton occupy the next trophic level. The next trophic levels are occupied by primary carnivores such as worms and molluscs and by large carnivore fishes.

Here also there is a gradual loss in the biomass from the basic trophic level to the higher tropic levels. In case of parasites like pyramid of numbers, the pyramid of biomass is inverted. Similarly, if organisms of lower levels average much smaller than those of higher levels, biomass pyramid may be inverted. For example, where size of the producers is very small and that of the consumers large, the total weight of the latter may be greater at any one moment. In such cases, even though more energy is being passed through the producer trophic level than through consumer levels, the rapid metabolism and turnover of the small producer organisms accomplish a larger output with a smaller standing crop biomass.

3. *Pyramid of Energy:* It indicates the total energy at each trophic level of the food chain. It also exhibits that at each trophic level loss of energy and material takes place as the processes of assimilation and growth are not 100 per cent efficient. Thus at the producer level, the total energy available is relatively more than at the higher trophic levels because of the loss of energy from one trophic level to the other. This means that the organic matter produced per average unit of time, and the energy represented by it, become less at each trophic level. The production rate of energy at different trophic levels of an ecosystem can be represented by the pyramid of energy. The base of such a pyramid is represented by the autotrophs, *i.e.* green plants and the higher levels are represented by different herbivore and carnivore trophic levels.

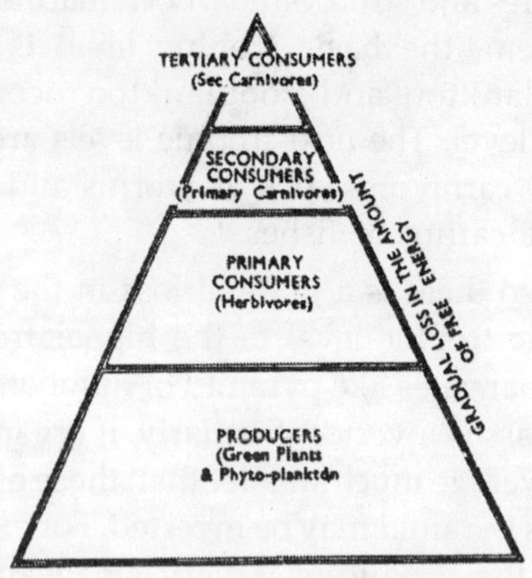

Pyramid of energy.

Principles of Ecosystem

Various kinds of life-supporting systems like the forests, grasslands, oceans, lakes, rivers, mountains, deserts and estuaries show wide variations in their structural composition and functions. However, they all are alike in the fact that they consist of living entities interacting with their surroundings exchanging matter and energy. How do these different units like a hot desert, a dense evergreen forest, the Antarctic Sea or a shallow pond differ in the type of their flora and fauna, how do they derive their energy and nutrients to live together, how do they influence each other and regulate their stability are the questions that are answered by ecology.

The term ecology was coined by Earnst Haeckel in 1869. It is derived from the Greek words *Oikos*- home + *logos*- study. So ecology deals with the study of organisms in their natural home interacting with their surroundings. The surroundings or environment consists of other living organisms (biotic) and physical (abiotic) components. Modern ecologists believe that an adequate definition of ecology must specify some unit of study and one such basic unit described by Tansley (1935) was ecosystem. An ecosystem is a group of biotic communities of species interacting with one another and with their non-living environment exchanging energy and matter. Now ecology is often defined as "the study of ecosystems".

An ecosystem is an integrated unit consisting of interacting plants, animals and microorganisms whose survival depends upon the maintenance and regulation of their biotic and abiotic structures and functions.

The ecosystem is thus, a unit or a system which is composed of a number of subunits, that are all directly or indirectly linked with each other. They may be freely exchanging energy and matter from outside—an open ecosystem or may be isolated from outside—a closed ecosystem.

PECULIARITIES OF ECOSYSTEM

Ecosystems show large variations in their size, structure, composition, etc. However, all the ecosystems are characterised by certain basic structural and functional features which are common.

Structural Features: Composition and organisation of biological communities and abiotic components constitute the structure of an ecosystem.

Biotic Structure: The plants, animals and microorganisms present in an ecosystem form the biotic component. These organisms have different nutritional behaviour and status in the ecosystems and are accordingly known as Producers or Consumers, based on how do they get their food.

(a) *Producers:* They are mainly the green plants, which can synthesise their food themselves by making use of carbon dioxide present in the air and water in the presence of sunlight by involving chlorophyll, the green pigment present in the leaves, through the process of photosynthesis. They are also known as photo autotrophs (auto=self; troph=food, photo=light).

There are some microorganisms also which can produce

organic matter to some extent through oxidation of certain chemicals in the absence of sunlight. They are known as chemosynthetic organisms or chemo-autotrophs.

For instance in the ocean depths, where there is no sunlight, chemoautotrophic sulphur bacteria make use of the heat generated by the decay of radioactive elements present in the earth's core and released in ocean's depths. They use this heat to convert dissolved hydrogen sulphide (H_2S) and carbon dioxide (CO_2) into organic compounds.

(b) *Consumers:* All organisms which get their organic food by feeding upon other organisms are called consumers, which are of the following types:

(i) *Herbivores (Plant Eaters):* They feed directly on producers and hence also known as primary consumers, *e.g.* rabbit, insect, man.

(ii) *Carnivores (Meat Eaters):* They feed on other consumers. If they feed on herbivores they are called secondary consumers (*e.g.* frog) and if they feed on other carnivores (snake, big fish, etc.) they are known as tertiary carnivores/consumers.

(iii) *Omnivores:* They feed on both plants and animals, *e.g.* humans, rat, fox, many birds.

(iv) *Detritivores (Detritus Feeders or Saprotrophs):* They feed on the parts of dead organisms, wastes of living organisms, their cast-offs and partially decomposed matter, *e.g.* beetles, termites, ants, crabs, earthworms, etc.

(c) *Decomposers:* They derive their nutrition by breaking down the complex organic molecules to simpler organic compounds and ultimately into inorganic nutrients. Various bacteria and fungi are decomposers.

In all the ecosystems, this biotic structure prevails. However, in some, it is the primary producers which predominate (*e.g.*

in forests, agro-ecosystems) while in others the decomposers predominate (*e.g.* deep ocean).

Abiotic Structure: The physical and chemical components of an ecosystem constitute its abiotic structure. It includes climatic factors, edaphic (soil) factors, geographical factors, energy, nutrients and toxic substances.

(a) *Physical Factors:* The sunlight and shade, intensity of solar flux, duration of sun hours, average temperature, maximum-minimum temperature, annual rainfall, wind, latitude and altitude, soil type, water availability, water currents, etc. are some of the important physical features which have a strong influence on the ecosystem.

We can clearly see the striking differences in solar flux, temperature and precipitation (rainfall, snow, etc.) pattern in a desert ecosystem, in a tropical rainforest and in tundra ecosystem.

(b) *Chemical Factors:* Availability of major essential nutrients like carbon, nitrogen, phosphorus, potassium, hydrogen, oxygen and sulphur, level of toxic substances, salts causing salinity and various organic substances present in the soil or water largely influence the functioning of the ecosystem.

All the biotic components of an ecosystem are influenced by the abiotic components and vice versa, and they are linked together through energy flow and matter cycling .

Elements Affecting Ecosystem

Every ecosystem performs under natural conditions in a systematic way. It receives energy from the sun and passes it on through various biotic components and in fact, all life depends upon this flow of energy. Besides energy, various nutrients and water are also required for life processes which are exchanged by the biotic components within themselves and with their abiotic components within or outside the ecosystem. The biotic components also regulate themselves in a very systematic manner and show mechanisms to encounter some degree of environmental stress.

The major functional attributes of an ecosystems are as follows:

(i) Food chain, food webs and trophic structure
(ii) Energy flow
(iii) Cycling of nutrients (Biogeochemical cycles)
(iv) Primary and Secondary production
(v) Ecosystem development and regulation

Trophic Structure: The structure and functions of ecosystems are very closely related and influence each other so intimately that they need to be studied together. The flow of energy is mediated through a series of feeding relationships in a definite sequence or pattern which is known as food chain. Nutrients too move along the food chain. The producers and consumers are arranged in the ecosystem in a definite manner and their interaction along with population size are expressed together as trophic structure. Each food level is known as trophic level and the amount of living matter at each trophic level at a given time is known as standing crop or standing Bio-mass.

Before we study about energy flow or nutrient cycling, we must learn about the food-chains, that provide the path through which the flow of energy and matter take place in ecosystem.

Sequential Structure of Food

The sequence of eating and being eaten in an ecosystem is known as food chain. All organisms, living or dead, are potential food for some other organism and thus, there is essentially no waste in the functioning of a natural ecosystem. A caterpillar eats a plant leaf, a sparrow eats the caterpillar, a cat or a hawk eats the sparrow and when they all die, they are all consumed by microorganisms like bacteria or fungi (decomposers) which break down the organic matter and convert it into simple inorganic substances that can again be used by the plants—the primary producers.

Some common examples of simple food chains are:

- Grass → grasshopper → Frog → Snake → Hawk (Grassland ecosystem)
- Phytoplanktons → water fleas → small fish → Tuna (Pond ecosystem)
- Lichens → reindeer → Man (Arctic tundra)

Each organism in the ecosystem is assigned a feeding level or trophic level depending on its nutritional status. Thus, in the grassland food chain, grasshopper occupies the 1st trophic level, frog the IInd and snake and hawk occupy the IIIrd and the IVth trophic levels, respectively. The decomposers consume the dead matter of all these trophic levels. In nature, we come across two major types of food chains:

I. *Grazing Food Chain:* It starts with green plants (primary producers) and culminates in carnivores. All the examples cited above show this type of food chain. Another example could be Grass → Rabbit → Fox

II. *Detritus Food Chain:* It starts with dead organic matter which the detritivores and decomposers consume. Partially decomposed dead organic matter and even the decomposers are consumed by detritivores and their predators. An example of the detritus food chain is seen in a mangrove (estuary).

Here, a large quantity of leaf material falls in the form of litter into the water. The leaf fragments are eaten by saprotrophs. (Saprotrophs are those organisms which feed on dead organic matter). These fallen leaves are colonised by small algae, which are also consumed by the saprotrophs or detritivores consisting of crabs, mollusks, shrimps, insect larvae, nematodes and fishes. The detritivores are eaten by small carnivorous fishes, which is turn are eaten by large carnivorous fishes.

Leaf litter → algae → crabs → small carnivorous fish → large carnivorous fish (mangrove ecosystem).

Dead organic matter → fungi → bacteria (forest ecosystem).

Thus the grazing food chain derives its energy basically from plant energy while in the detritus food chain it is obtained primarily from plant Bio-mass, secondarily from microbial Bio-mass and tertiarily from carnivores. Both the food chains occur together in natural ecosystems, but grazing food chain usually predominates.

Food Web: Food chains in ecosystems are rarely found to operate as isolated linear sequences. Rather, they are found to be interconnected and usually form a complex network with several linkages and are known as food webs. Thus, food web is a network of food chains where different types of organisms are connected at different trophic levels, so that there are a number of options of eating and being eaten at each trophic level. A food-web in the unique Antarctic Ecosystem is representing the total ecosystem including the Antarctic sea and the Continental land. The land does not show any higher life forms of plants. The only species are that of some algae, lichens and mosses. The animals include penguins and snow petrel which depend upon the aquatic chain for their food energy.

In a tropical region, on the other hand, the ecosystems are much more complex. They have a rich species diversity and therefore, the food webs are much more complex.

Why nature has evolved food webs in ecosystems instead of simple linear food chains? This is because food webs give greater stability to the ecosystem. In a linear food chain, if one species becomes extinct or one species suffers, the species in the subsequent trophic levels are also affected. In a food web, on the other hand, there are a number of options available at each trophic level. So if one species is affected, it does not affect other trophic levels so seriously.

Just consider the simple food chains of arctic tundra

ecosystem: Cladonia → Reindeer → Man Grass → Caribou → Wolf

If due to some stress, the population of reindeer or Caribou falls, it will leave little option for man or wolf to eat from the ecosystem. Had there been more biodiversity, it would have led to complex food web giving the ecosystem more stability.

Significance of Food Chains and Food Webs: Food chains and food webs play a very significant role in the ecosystem because the two most important functions of energy flow and nutrient cycling take place through them.

The food chains also help in maintaining and regulating the population size of different animals and thus, help maintain the ecological balance.

Food chains show a unique property of biological magnification of some chemicals. There are several pesticides, heavy metals and other chemicals which are non-biodegradable in nature. Such chemicals are not decomposed by microorganisms and they keep on passing from one trophic level to another. And, at each successive trophic level, they keep on increasing in concentration. This phenomenon is known as biomagnification or biological magnification.

Pyramids in Ecological System: Graphic representation of trophic structure and function of an ecosystem, starting with producers at the base and successive trophic levels forming the apex is knows as an ecological pyramid. Ecological pyramids are of three types:

Pyramid of Numbers: It represents the number of individual organisms at each trophic level. We may have upright or inverted pyramid of numbers, depending upon the type of ecosystem and food chain. A grassland ecosystem and a pond ecosystem can have upright pyramid of numbers. The producers in the grasslands are grasses and that in a pond are phytoplanktons (algae, etc.), which are small in size and very large in number. So the producers form a broad base. The herbivores in a grassland are insects while tertiary carnivores are hawks or other birds which are gradually less and less in number and hence the pyramid apex becomes gradually narrower forming an upright pyramid. Similar is the case with the herbivores, carnivores and top carnivores in pond which decrease in number at higher trophic levels.

In a forest ecosystem, big trees are the producers, which are less in number and hence form a narrow base. A larger number of herbivores including birds, insects and several species of animals feed upon the trees (on leaves, fruits, flowers, bark, etc.) and form a much broader middle level. The secondary consumers like fox, snakes, lizards, etc. are less in number than herbivores while top carnivores like lion, tiger, etc. are still smaller in number. So the pyramid is narrow on both sides and broader in the middle.

Parasitic food chain shows an inverted pyramid of number. The producers like a few big trees harbour fruit, eating birds acting like herbivores which are larger in number. A much higher number of lice, bugs, etc. grow as parasites on these birds while a still greater number of hyperparasites like bugs, fleas and microbes feed upon them, thus making an inverted pyramid.

Pyramid of Bio-mass: It is based upon the total Bio-mass (dry matter) at each trophic level in a food chain. The pyramid of Bio-mass can also be upright or inverted. The pyramid of Bio-mass in a forest is upright in contrast to its pyramid of numbers. This is because the producers (trees) accumulate a huge biomass while the consumers' total biomass feeding on them declines at higher trophic levels, resulting in broad base and narrowing top.

The pond ecosystem shows an inverted pyramid of biomass. The total biomass of producers (phytoplanktons) is much less as compared to herbivores (zooplanktons, insects), Carnivores (Small fish) and tertiary carnivores (big fish). Thus the pyramid takes an inverted shape with narrow base and broad apex.

Pyramid of Energy: The amount of energy present at each trophic level is considered for this type of pyramid. Pyramid of energy gives the best representation of the trophic relationships and it is always upright.

At every successive trophic level, there is a huge loss of energy (about 90%) in the form of heat, respiration, etc. Thus, at each next higher level only 10 per cent of the energy passes on. Hence, there is a sharp decline in energy level of each successive trophic level as we move from producers to top carnivores. Therefore, the pyramid of energy is always upright.

Flow of Energy in an Ecosystem: Flow of energy in an ecosystem takes place through the food chain and it is this energy flow which keeps the ecosystem going. The most important feature of this energy flow is that it is unidirectional or oneway flow. Unlike the nutrients (like carbon, nitrogen, phosphorus, etc.) which move in a cyclic manner and are reused by the producers after flowing through the food chain, energy is not reused in the food chain. Also, the flow of energy follows the two laws of Thermodynamics.

First law of Thermodynamics states that energy can neither be created nor be destroyed but it can be transformed from one form to another. The solar energy captured by the green plants (producers) gets converted into biochemical energy of plants and later into that of consumers.

Dud law of Thermodynamics states that energy dissipates as it is used or in other words, its gets converted from a more concentrated to dispersed form. As energy flows through the food chain, there occurs dissipation of energy at every trophic level. The loss of energy takes place through respiration, loss of energy in locomotion, running, hunting and other activities. At every level there is about 90 per cent loss of energy and the energy transferred from one trophic level to the other is only about 10 per cent.

Energy Flow Models: The flow of energy through various trophic levels in an ecosystem can be explained with the help of various energy flow models.

(a) *Universal Energy Flow Model:* Energy flow through an ecosystem was explained by E.P. Odum as the universal energy flow model. As the flow of energy takes place, there is a gradual loss of energy at every level, thereby resuiting in less energy available at next trophic level as indicated by narrower pipes (energy flow) and smaller boxes (stored energy in biomass). The loss of energy is mainly the energy not utilised (NU). This is the energy lost in locomotion, excretion, etc. or it is the energy lost in respiration (R) which is for maintenance. The rest of the energy is used for production (P).

(b) *Single Channel Energy Flow Model:* The flow of energy takes place in an unidirectional manner through a single channel of green plants or producers to herbivores and carnivores.

(c) *Double Channel or Y-shaped Energy Flow Model:* In nature, both grazing food chain and detritus food chain operate in the same ecosystem. However, sometimes it is the grazing food chain which predominates. It happens in marine ecosystem where primary production in the open sea is limited and a major portion of it is eaten by herbivorous marine animals. Therefore, very little primary production is left to be passed on to the dead or detritus compartment. On the other hand, in a forest ecosystem the huge quantity of biomass produced cannot be all consumed by herbivores. Rather, a large proportion of the live biomass enters into detritus (dead) compartment in the form of litter. Hence the detritus food chain is more important there.

The two channel or Y-shaped model of energy flow shows the passage of energy through these two chains, which are separated in time and space.

Function of Nutrient Cycling: Besides energy flow, the other important functional attribute of an ecosystem is nutrient cycling. Nutrients like carbon, nitrogen, sulphur, oxygen, hydrogen, phosphorus, etc. move in circular paths through biotic and abiotic components and are therefore known as biogeochemical cycles. Water also moves in a cycle, known as hydrological cycle. The nutrients too move through the food chain and ultimately reach the detritus compartment (containing dead organic matter) where various micro-organisms carry out decomposition. Various organically, bound nutrients of dead plants and animals are converted into inorganic substances by microbial decomposition that are readily used up by plants (primary producers) and the cycle starts afresh.

Function of Nitrogen Cycle: Nitrogen is present in the atmosphere as N_2 in large amount (78%) and it is fixed either by the physical process of lightening or biologically by some bacteria and/or cyanobacteria (blue green algae). The nitrogen is taken up by plants and used in metabolism for biosynthesis of amino-acids, proteins, vitamins, etc. and passes through the food chain. After death of the plants and animals, the organic nitrogen in dead tissues is decomposed by several groups of ammonifying and nitrifying bacteria which convert them into ammonia, nitrites and nitrates, which are again used by plants. Some bacteria convert nitrates, into molecular nitrogen or N_2 which is released back into the atmosphere and the cycle goes on.

Function of Carbon Cycle: Sometimes human interferences disturb the normal cycling of such nutrients and create imbalances. For example, nature has a very balanced carbon cycle. Carbon, in the form of carbon dioxide is taken up by green plants as a raw material for photosynthesis, through which a variety of carbohydrates and other organic substances are produced. Through the food chain it moves and ultimately organic carbon present in the dead matter is returned to the atmosphere as carbon dioxide by microorganisms. Respiration by all organisms produces carbon dioxide, while the latter is used up by plants.

In the recent years carbon dioxide levels have increased in the atmosphere due to burning of fossil fuels, etc. which has caused an imbalance in the natural cycle and the world today is facing the serious problem of global warming due to enhanced carbon dioxide emissions.

Function of Phosphorus Cycle: Phosphorous cycle is another important nutrient cycle. The reservoir of phosphorus lies in the rocks, fossils, etc. which is excavated by man for using it as a fertilizer. Farmers use the phosphate fertilizers indiscriminately and as a result excess phosphates are lost as runoff, which causes the problem of eutrophication or overnourishment of lakes leading to algal blooms. A good proportion of phosphates moving with surface runoff reaches the oceans and are lost into the deep sediments.

Our limited supply of phosphorus lying in the phosphate rocks of this earth are thus overexploited by man and a large part is taken out of the normal cycle due to loss into oceans. So human beings are making the phosphorous cycle acyclic. Sea birds, on the other hand, are playing an important role in phosphorus cycling.

They eat sea-fishes which are phosphorus-rich and the droppings or excreta of the birds return the phosphorus on the land. The Guano deposits on the coasts of Peru are very rich sources of phosphorus.

Production in Primary Stage: Primary productivity of an ecosystem is defined as the rate at which radiant energy is converted into organic substances by photosynthesis or chemosynthesis by the primary producers.

When organic matter is produced by the primary producers (mainly green plants and some microorganisms), some of it is oxidized or burnt inside their body and converted into carbon dioxide which is released during respiration and is accompanied by loss of energy.

Respiratory loss of energy is a must, because it is required for the maintenance of the organism. Now, the producers are left with a little less organic matter than what was actually produced by them.

This is known as the Net Primary Production (NPP) and the respiratory loss (R) added to it gives the Gross Primary Production (GPP).

Thus, NPP = GPP – R.

Primary production of an ecosystem depends upon the solar radiations, availability of water and nutrients and upon the type of the plants and their chlorophyll content. Below given table shows the average gross primary productivity of some major ecosystems.

Annual Average of Gross Primary Production of Some Major Ecosystems

Ecosystem	*Gross Primary Productivity (K Cal/m^2/yr)*
Deserts and Tundra	200
Open Oceans	1,000
Grasslands	2,500
Moist Temperate Forests	8,000
Agro-ecosystems	12,000
Wet Tropical Forests	20,000
Estuaries	20,000

Productivity of tropical forests and estuaries are the highest. This is because tropical forests have abundant rainfall, warm temperature congenial for growth, abundant sunlight and a rich diversity of species. Estuaries get natural energy subsidies in the form of wave currents that bring along with them nutrients required for production. Deserts, on the other hand, have limitations of adequate water supply while Tundra have very low temperature as limiting factor and hence show low primary production.

Agro-ecosystems get lots of energy subsidies in the form of irrigation water, good quality seeds, fertilizers and pesticides and show a high productivity of 12,000 K Cal/m^2/yr. Still, it is noteworthy that their productivity is less than that of tropical forests which are not receiving any artificial energy subsidies.

Nature itself has designed its species composition, structure, energy capture and flow, and a closed nutrient cycling system that ensures a high primary production of 20,000 K Cal/m^2/yr. Also, the qualitative variety of the primary production is enormous in the tropical forests. This makes it all the more important to conserve our tropical forests.

Production in Secondary Stage: The food synthesised by green plants through photosynthesis is the primary production which is eaten by herbivores. The plant energy is used up for producing organic matter of the herbivores which, in turn, is used up by the carnivores. The amount of organic matter stored by the herbivores or carnivores (in excess of respiratory loss) is known as secondary production. The energy stored at consumer level for use by the next trophic level is thus defined as secondary production.

Regulating Ecosystem: All ecosystems regulate themselves and maintain themselves under a set of environmental conditions. Any environmental stress tries to disturb the normal ecosystem functions. However, the ecosystem, by itself, tries to resist the change and maintain itself in equilibrium with the environment due to a property known as homeostasis. Homeostasis is the inherent property of all living systems to resist change. However, the system can show this tolerance or resistance only within a maximum and a minimum range, which is its range of tolerance known as homeostatic plateau. Within this range, if any stress tries to cause a deviation, then the system has its own mechanisms to counteract these deviations which are known as negative feedback mechanisms. So negative feedback mechanisms are deviation counteracting mechanisms which try to bring the system back to its ideal conditions.

But, if the stress is too high and beyond the range of homeostatic plateau, then another type of mechanisms known as positive feedback mechanisms start operating. These are the deviation accelerating mechanisms. So the positive feedback mechanisms add to the stress conditions and tend to take the system away from the optimal conditions. Human beings should try to keep the ecosystems within the homeostatic plateau. They should not contribute to positive feedbacks. Otherwise the ecosystems will collapse.

Food Cycle

As a matter of fact the food chains are not isolated sequences but are interconnected with one another. For example, man and many other organisms are at the same time herbivore and carnivore and hence occupy different trophic levels in the food chain. There may be several carnivores also with overlapping prey types. This trophic relationship between organisms of an ecosystem is not always in a simple chain-like fashion but forms a complicated network. Furthermore, different types of food chains interact upon each other. This interlocking pattern of several food chains is known as food web. Thus food web may be defined as the relationship in which a predator eats several types of food and every kind of food is eaten by many different organisms and is the outcome of the interaction between different types of food chains.

Trophic Levels and Ecological Pyramids

Trophic Levels: Each successive level of nourishment as represented by the links of the food chain is known as a trophic level. The plant producers within an ecosystem constitute the first trophic level, the herbivores from the second trophic level, and the carnivores represent third levies. Additional links in the main food chain and inside chains such as those formed by parasites constitute further trophic levels.

Step-wise Transfer of Energy

The existence of living world depends upon the flow of energy and circulation of materials through the ecosystem. The energy is required for the performance of all the life activities. The source of this energy is sun. The solar energy enters the space in the form of light rays. Approximately 57 per cent of solar energy is absorbed in the atmosphere and scattered in space. About 36 per cent is expended in heating water and land and in evaporating water. Nearly 8 per cent of light energy strikes the plants, of which 80-85 per cent is absorbed, and only fifty per cent of it is utilised in photosynthesis.

This energy is captured by plants and is stored in the form of potential energy in foodstuffs. These are known as producers and represent the first trophic level in the ecosystem. The energy stored by plants is passed through the community or ecosystem in a series of steps of eating and being eaten. This is known as a food chain. A food chain consists of four steps- the producers, primary consumers, secondary consumers and tertiary consumers. The energy flows from the producers to consumers. At each transfer a large proportion (80 to 90 per cent) of potential energy is dissipated as heat produced during the process of respiration and other ways.

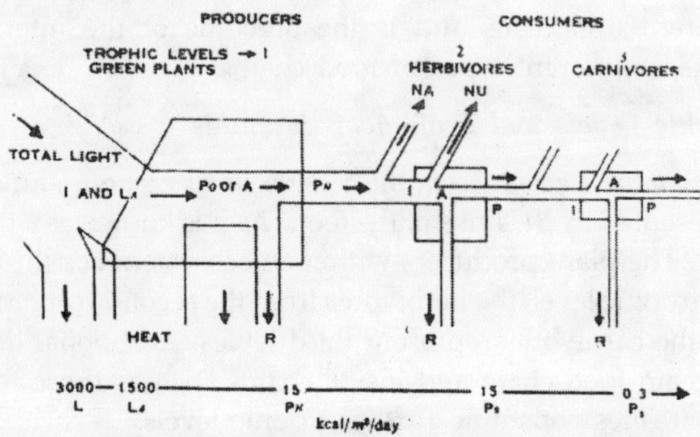

Diagrammatic representation of energy flow through a food chain of ecosystem. The boxes represent biomass or population mass and the pipes show the path of flow of energy between living units. The relative size of block suggests the quantity of energy flowing through each pipe. L = total energy input; LA = light absorbed by plants; PG = Primary groos production; A =total assimilation; PN = net primary production; P = secondary production; NU = energy not used; NA = energy not assimilated by consumers; R=respiration

The energy flow through an ecosystem can be represented diagrammatically in a simplified manner. In the boxes represent the trophic levels and the pipes depict the energy flow in and out at each level. Only about half the average light energy impinging upon the green plants is absorbed in the photosynthetic machinery, out of which 1 to 5 per cent is converted into food energy and the rest of it passes out as heat in the atmosphere. Energy accumulated by plants or the producers in an ecosystem is called primary production.

The total energy produced during photosynthesis is the groos primary production and is represented by PG or A and the energy left after respiration and stored as organic matter in the producers is the net primary production represented by PN. Net primary consumers, actually represent food potentially available to primary production which feed upon plants. The primary consumers, therefore, take in chemical potential energy in the form of plant food. Most of it dissipates in the form of heat (produced during respiration) and is lost out of the ecosystem.

Only a small part of energy is fixed in the form of chemical potential energy in the protoplasm. The same process is repeated at the secondary consumers or primary carnivorous levels and so on. Therefore, at each step in the transfer of energy from one trophic level to another a larger amount of energy is degraded into heat and never returns to the ecosystem.

In the transformation of energy through ecosystem, the energy is reduced in magnitude by 100 from primary producers to plant consumers and by 10 for each step thereafter.

Primary productivity of an ecosystem or community is defined as the rate at which radiant energy is stored by photosynthetic and chemosynthetic activities of producer organisms (chiefly green plants) in the form of organic substances which can be used as food materials. There are four successive steps in the production process as follows:

1. *Groos Primary Productivity:* It is the total energy stored

in the food materials synthesised by the green plants or total rate of photosynthesis including the organic matter used up in the respiration during "total photosynthesis or total assimilation."

2. *Net Primary Productivity:* It is the total gross productivity minus the energy used up in the metabolic process. In other words it is the rate of storage of organic matter in plants in excess of the organic matter utilised in respiration by the plants during the measurement period. This is also called apparent photosynthesis or net assimilation.
3. *Net Community Productivity:* It is the rate of storage of organic matter not used by heterotrophs during the period under consideration, usually the growing season or a year. This is net primary production minus heterotophic consumption.
4. *Secondary Productivities:* The rate of energy storage at consumer levels are referred to as secondary productivities.

There are three fundamental aspects of productivity: (1) standing crop, (2) material removed and (3) production rate.

Complete Crops

The amount of living material in different trophic levels or in a component population is known as the standing crop. The standing crop can be expressed in terms of the number per unit area or in terms of biomass, that is organism's mass. Biomass can be measured as living weight, dry weight, ash-free dry weight, carbon weight, calories or any other unit that may be useful for comparative purposes.

The standing crop not only represents potential energy, but is also important as a habitat or living space for organisms. Thus the trees in a forest not only represent energy that provides food or fuel but they modify climate and provide shelter for birds and other animals.

Expulsion of Material

The second concept of productivity is the material removed from the area per unit time, and includes the yield to man, organisms removed from the ecosystem by emigration and material withdrawn as organic deposits.

For a community to maintain itself, material in some form must be supplied to it in an amount equal to the amount of material removed from the area. The supply in the form of nutrients and immigrants usually should be considerably larger than the material removed because of inevitable loss of energy and of material from the system at each level in the food chain.

The concept of material removed can be illustrated well by considering a natural pasture from which it is desired to obtain the maximal yield of sheep. Increased growth of the plant producers will result by the addition of fertilizers. The grass supports not only a flock of sheep but it also supports a population of field rodents. If the rodents are excessively abundant, a reduction in their number would allow a larger portion of the grass for the sheep.

The rodents could be controlled by allowing the presence of predators, but should the predators become too numerous, they would compete with man for the sheep, and thus their number might have to be reduced. A suitable balance is to be maintained among all the members of the ecosystem. The population of the species to be harvested should be adjusted to the level of abundance so that the growth rate is as great as possible without disturbing the equilibrium of the community. The maximal potential is the large maintained harvest that can be removed from an area under the best population conditions.

The Productivity

The production rate is the rate at which growth processes are going forward within the area. In areas in which rates of production are higher than the rates of destruction, a net increase in one or more of the components will occur. The surplus material may be removed by man, which constitutes a yield, or it may be removed in other ways or may be permanently added to the system. In the last case an increase in the organic components of the ecosystem will be experienced, but such an increase depends on the amount of nutrients added to the area. The amount of organisms and of nutrients in a lake, for example, are frequently seen to increase in this way over a period of time, with the result that the lakes become more eutrophic and the process is known as eutrophication. Similar increase of the living and nonliving components of ecosystem occurs in other types of habitats.

Functions of Homoeostasis

Changes which occur in ecosystem over a geological time span are met by evolutionary changes resulting from natural selection. Those changes that occur naturally as a result of normal succession are met by changes m species makeup of the communities involved. The ecosystem always makes of most efficient use of the available energy within the limitations of the physical environment. Whether naturally or artificially caused, changes are met by adjustment of the species present or the creation and filling of new ecological niches by species which are new to the ecosystem. At population and community levels organisation, controls result from the many intraspecific and interspecific relationships.

Water in Ecosystem

Of all the ecosystems, ocean is the largest and most stable ecosystem. Aquatic life is protected from vigorous climates and weather that are faced by the organisms of terrestrial environment. The unpredictable climatic conditions, problem of water supply, food, fire and artificial forces such as industrialisation, farming and grazing are lacking in the oceans. A simple ocean ecosystem may be described as follows:

1. *Primary Producers:* Planktonic plants are the producers

of the open sea. These are diatoms, dinoflagellates, few microscopic and unicellular algae, weeds and brown and red algae. These form the basis of all other life found in sea. These present upto the depth to which the light penetrates.

2. *Consumers:* The primary consumers or herbivores of the sea are crustaceans, molluscs, certain sessile invertebrates and herbivorous fishes.

The secondary consumers or the primary carnivores of the sea are certain large-sized fish such as mackerel, herring, basking shark, etc. The secondary carnivores in the mature food chain are fishes such as cod, haddock, halibut and flounders. These are bottom feeders preying upon worms, molluscs, crabs and other benthic animals. Though certain molluscs are primarily herbivores, but other types of benthos are secondary or tertiary carnivores, scavengers or mud feeders. Many worms and echinoderms pass large amount of mud through their intestines, extracting nutrient from its components. Certain fish feed primarily upon smaller members of their own class and the growth of shark represent a food chain of five or more links.

Decomposers and Transformers: The decomposers of sea are marine bacteria. These decompose the excreta of living organisms and the dead and decaying organisms. The soluble and decomposed material is now attached by other types of bacteria, the transformers, and converted to forms to be reused by the green plants.

Meadow in Ecosystem

Grasslands occupy about 19 per cent of the earth's surface. The major grassland ecosystems of the world are the great plains of Canada and United States, S. Argentina to Brazil and S. Asia to Central Asia. The various components of a grassland ecosystem are as follows:

1. *Abiotic Substances:* These include the nutrients present in the soil and the aerial environment. The elements

required by plants are hydrogen, oxygen, nitrogen, phosphorous and sulphur. These are supplied by the soil and air in the form of CO_2, water, nitrates, phosphates and sulphates. In addition to these some trace elements are also present in the soil.

2. *Primary Producers:* These are mainly grasses of the family Graminae, a large variety of herbs, some shrubs and scattered trees.
3. *Consumers:* Herbivores such as grazing mammals (*e.g.* cows, sheep, deer, rabbit, buffaloes, etc.), insects (*e.g.* Dysdercus, Coccinella, Leptocorisa, etc.), some termites and millipedes are the primary consumers.

   The animals like fox, jackals, snakes, frogs, lizards, birds, etc. are the carnivores feeding on herbivores. These are the secondary consumers of the grassland ecosystem. Hawks occupy the tertiary trophic level as these feed on the secondary consumers.
4. *Decomposers:* These include bacteria of death and decay, moulds and fungi (*e.g.* Mucor, Penicillium, Aspergillus, Rhizopus, etc.) These bring the minerals back to the soil to be available to the producers again.

Ecosystem and Soil

This is a man made ecosystem. Man has been doing his best in modifying the croplands to get maximum benefit out of them. A cropland ecosystem may be illustrated by crops like wheat, maize, rice, sugarcane, etc. It has following components:

1. *Producers:* A crop along with the weeds growing in the field are the producers of a cropland ecosystem. In the cropland ecosystem of maize weeds like Cynodon dactylon, Euphorbia hirta, Alysicarpus, Launea nudicaulis are very common.
2. *Consumers:* Herbivores one the primary consumers of the cropland ecosystem. These include rats, rabbits, birds, man and insects. The second order and the third

order consumers are represented by frogs, anakes, birds such as hawk, etc.

3. *Decomposers and Transformers:* Bacteria and fungi like Bacillus, Aspergillus, Clostridium, Agaricns, Mucor, Aspergillns and Fusarium are common decomposers and transformers of cropland ecosystem of maize.

Forest being a Part of Ecosystem

Forests occupy roughly 40 per cent of the total land. In India about 1/10th of the total area is under forest. Depending on climatic conditions the forests have been divided into four categories:

1. Tropical forests
2. Subtropical forests
3. Temperate forests
4. Alpine forests.

All the above types of forests have their own typical fauna and flora. The chief components of a forest ecosystem are as follows:

1. *Abiotic Substances:* In addition to minerals present in the soil, the forest floor is rich in dead and decaying organic matter.
2. *Primary Producers:* The dominant producers of the forest ecosystem are the higher forms of the plant kingdom, most specifically seed bearing plants. In forest ecosystem the producers form three major functions. First, they provide the initial source of food. Second, large plant structures provide habitats for other organisms. Third, they are prime agents in soil formation and in modifying the abiotic environment.

The forest ecosystem is directly dependent upon the kinds and numbers of plants present. The plants are mainly trees, some shrubs and ground vegetation.

The evergreen tropical forests have broad leaved trees

with irregular leaf fall. The temperate deciduous forests are dominated by pines with needle-like leaves. At the lower level of such forests are found only shade tolerant plants such as ferns and shrubs. The other forms of tropical evergreen forests are the vines and the epiphytes.

3. *Primary Consumers:* These are the herbivores that include the foliage arthropods such as ants, flies, beetles, leaf-hoppers, bugs and spiders, etc. and the larger animals grazing on shoots and fruits such as elephants, nilgai, deer, moles, squirrels, shrews, flying foxes, fruit bats and monogooses, etc.
4. *Secondary Consumers:* These are the carnivores like snakes, birds, lizards and foxes feeding on herbivores. The top carnivores like lion, tiger, etc. prey upon both herbivores and carnivores of the secondary consumer level.
5. *Decomposers:* The decomposers of the forest ecosystem, as in other ecosystem, are the bacteria and fungi. These organisms obtain energy to carry on the their life functions by breaking down the organic compounds of dead organic matter and organic wastes. In doing so, they prevent the accumulation of organic matter and also release the essential minerals for reuse. Decomposers are mostly found in soil and hetertrophic bacteria are by the numerous.

Desert being a part of Ecosystem

The deserts occupy about 17 per cent of the land and occur in the regions with an average rainfall of less than 23 cms. Due to extremes of temperature, the species composition of desert ecosystem is much varied and typical. The various components of a desert ecosystem are:

1. *Producers:* The shrubs, bushes, grasses and some trees are the main producers in deserts. The shrubs have extensive and much branched root system with the

stems and leaves variously modified. Some succulent cacti are also found in deserts. These store water in their stem to be used during the time of water scarcity. Some lower plants such as lichens, xerophytic mosses and blue green algae are also found there.

2. *Consumers:* Only a few animals are found in deserts. The most common animals are those reptiles and insects which are able to live under xeric conditions. Mammals are represented by a few species of nocturnal rodents. Some birds are also present. The camel, called the ship of desert, feeds on tender shoots of the plants and is to conserve large quantities of water in its stomach. The larger animals including carnivores are scarce. The desert animals have various morphological and physiological adaptations which enable them to live in such extreme environment.
3. *Decomposers:* Due to poor vegetation and less amount of dead organic matter, decomposers are few. They are thermophilic fungi and bacteria.

5

Natural Resources in Use

Resources from Nature

Life on this planet earth depends upon a large number of things and services provided by the nature, which are known as Natural resources. Thus water, air, soil, minerals, coal, forests, crops and wildlife are all examples of natural resources.

The natural resources are of two kinds:

- Renewable resources which are in-exhaustive and can be regenerated within a given span of time e.g. forests, wildlife, wind energy, Bio-mass energy, tidal energy, hydro power, etc. Solar energy is also a renewable form of energy as it is an inexhaustible source of energy.
- Non-renewable resources which cannot be regenerated e.g. Fossil fuels like coal, petroleum, minerals, etc. Once we exhaust these reserves, the same cannot be replenished.

Even our renewable resources can become non-renewable if we exploit them to such extent that their rate of consumption exceeds their rate of regeneration. For example, if a species is exploited so much that its population size declines below the threshold level then it is not able to sustain itself and gradually the species becomes endangered or extinct. It is very important to protect and conserve our natural resources and use them in a judicious manner so that we don't exhaust them. It does not mean that we should stop using most of the natural resources. Rather, we should use the resources in such a way that we always save enough of them for our future generations. In this unit we shall discuss the major natural resources: (i) Forest resources, (ii) Water resources, (iii) Mineral resources, (iv) Food resources, (v) Energy resources, (vi) Land resources.

Various Aspects of Forests

Forests are one of the most important natural resources on this earth. Covering the earth like a green blanket these forests not only produce innumerable material goods, but also provide several environmental services which are essential for life.

About l/3rd of the world's land area is forested which includes closed as well as open forests. Former USSR accounts for about a 5th of the world's forests, Brazil for about a 7th and Canada and USA each for 6-7%. But it is a matter of concern that almost everywhere the cover of the natural forests has declined over the years. The greatest loss occurred in tropical Asia where one third of the forest resources have been destroyed.

Jungles as Resources

Commercial uses: Forests provide us a large number of commercial goods which include timber, firewood, pulpwood, food items, gum, resins, non-edible oils, rubber, fibers, lac, bamboo canes, fodder, medicine, drugs and many more items, the total worth of which is estimated to be more than $ 300 billion per year.

Half of the timber cut each year is used as fuel for heating and cooking. One third of the wood harvest is used for building materials as lumber, plywood and hardwood, particle board and chipboard. One sixth of the wood harvest is converted into pulp and used for paper industry. Many forest lands are used for mining, agriculture, grazing, and recreation and for development of dams.

Ecological uses: While a typical tree produces commercial goods worth about $ 590 it provides environmental services worth nearly $ 196, 250. The ecological services provided by our forests may be summed up as follows:

- *Production of Oxygen:* The trees produce oxygen by photo synthesis which is so vital for life on this earth. They are rightly called as earth's lungs.
- *Reducing Global Warming:* The main greenhouse gas carbon dioxide (CO_2) is absorbed by the forests as a raw material for photosynthesis. Thus forest canopy acts as a sink for CO_2 thereby reducing the problem of global warming caused by greenhouse gas CO_2.
- *Wildlife Habitat:* Forests are the homes of millions of wild animals and plants. About 7 million species are found in the tropical forests alone.
- *Regulation of Hydrological Cycle:* Forested watersheds act like giant sponges, absorbing the rainfall, slowing down the runoff and slowly releasing the water for recharge of springs. About 50-80 %of the moisture in the air above tropical forests comes from their transpiration which helps in bringing rains.
- *Soil Conservation:* Forests bind the soil particles tightly in their roots and prevent soil erosion. They also act as wind breaks.
- *Pollution Moderators:* Forests can absorb many toxic gases and can help in keeping the air pure. They have also been reported to absorb noise and thus help in preventing air and noise pollution.

Since time immemorial, humans have depended heavily on forests for food, medicine, shelter, wood and fuel. With growing civilisation the demands for raw material like timber, pulp, minerals, fuel wood, etc. shooted up resulting in large scale logging, mining, road-building and clearing of forests. Our forests contribute substantially to the national economy. The international timber trade alone is worth over US $ 40 billion per year. Excessive use of fuel wood and charcoal, expansion of urban, agricultural and industrial areas and overgrazing have together led to over-exploitation of our forests leading to their rapid degradation.

Deforestation: The total forest area of the world in 1900 was estimated to be 7,000 million hectares which was reduced to 2890 million ha in 1975 and fell down to just 2,300 million ha by 2000. Deforestation rate is relatively less in temperate countries, but it is very alarming in tropical countries where it is as high as 40-50 per cent and at the present rate it is estimated that in the next 60 years we would lose more than 90 per cent of our tropical forests.

The forested area in India seems to have stabilised since 1982 with about 0.04% decline annually between 1982-90. FAO (1983) estimated that about 1.44 m ha of land was brought under afforestation during this period leading to stabilisation. As per FAO estimates, the deforestation rate per unit population in India is the lowest amongst the major tropical countries, despite the fact that we have a huge population size and very low per capita forest area (0.075 ha per capita). However, we are still far behind the target of achieving 33% forest area, as per our National Forest Policy, as we are still having only 19.27 % of our land area (63.38m ha), covered by forests based on satellite data (MoEF, 1998)

Principal Reasons behind Deforestation

(i) *Shifting Cultivation:* There are an estimated 300 million people living as shifting cultivators who practice slash and burn agriculture and are supposed to clear more than 5 lakh hectares of forests for shifting cultivation annually. In India, we have this practice in North-East and to some extent in Andhra Pradesh, Bihar and M.P. which contribute to nearly half of the forest clearing annually.

(ii) *Fuel Requirements:* Increasing demands for fuel wood by the growing population in India alone has shooted up to 300-500 million tons in 2001 as compared to just 65 million tons during independence, thereby increasing the pressure on forests.

(iii) *Raw Materials for Industrial Use:* Wood for making boxes, furniture, railway-sleepers, plywood, match-boxes, pulp for paper industry, etc. have exerted tremendous pressure on forests. Plywood is in great demand for packing tea for Tea industry of Assam while fir tree wood is exploited greatly for packing apples in J & K.

(iv) *Development Projects:* Massive destruction of forests occur for various development projects like hydroelectric projects, big dams, road construction, mining, etc.

(v) *Growing Food Needs:* In developing countries this is the main reason for deforestation. To meet the demands of rapidly growing population, agricultural lands and settlements are created permanently by clearing forests.

(vi) *Overgrazing:* The poor in the tropics mainly rely on wood as a source of fuel leading to loss of tree cover and the cleared lands are turned into the grazing lands. Overgrazing by the cattle leads to further degradation of these lands.

Principal Results of Deforestation: Deforestation has far reaching consequences, which may be outlined as follows:

(i) It threatens the existence of many wildlife species due to destruction of their natural habitat.

(ii) Biodiversity is lost and along with that genetic diversity is eroded.

(iii) Hydrological cycle gets affected, thereby influencing rainfall.

(iv) Problems of soil erosion and loss of soil fertility increase.

(v) In hilly areas it often leads to landslides.

Tasks of Forests

Timber Extraction: Logging for valuable timber, such as teak and Mahogany not only involves a few large trees per hectare but about a dozen more trees since they are strongly interlocked with each other by vines, etc. Also road construction for making approach to the trees causes further damage to the forests.

***Mining*:** Mining operations for extracting minerals and fossil fuels like coal often involves vast forest areas. Mining from shallow deposits is done by surface mining while that from deep deposits is done by sub-surface mining. More than 80,000 ha of land of the country is presently under the stress of mining activities. Mining and its associated activities require removal of vegetation along with underlying soil mantle and overlying rock masses. This results in defacing the topography and destruction of the landscape in the area.

Large scale deforestation has been reported in Mussorie and Dehradun valley due to indiscriminate mining of various minerals over a length of about 40 Km. The forested area has declined at an average rate of 33% and the increase in non-forest area due to mining activities has resulted in relatively unstable zones leading to landslides.

Indiscriminate mining in forests of Goa since 1961 has destroyed more than 50,000 ha of forest land. Coal mining in Jharia, Raniganj and Singrauli areas have caused extensive deforestation in Jharkhand. Mining of magnesite and soap-stones have destroyed 14 ha of forest in the hill slopes at Khirakot, Kosi valley, Almora.

Mining of radioactive minerals in Kerala, Tamil Nadu and Karnataka are posing similar threats of deforestation. The rich forests of Western Ghats are also facing the same threat due to mining projects for excavation of copper, chromite, bauxite and magnetite.

How Dams have its Imparts on Forests and People?: Big dams and river valley projects have multi-purpose uses and have been referred to as *"Temples of modern India"*. However, these dams are also responsible for the destruction of vast areas of forests.

India has more than 1550 large dams, the maximum being in the state of Maharashtra (more then 600), followed by Gujarat (more then 250) and Madhya Pradesh (130). The highest one is *Tehri dam,* on river Bhagirathi in Uttaranchal and the largest in terms of capacity is Bhakra dam on river Satluj in H.P. Big dams have been in sharp focus of various environmental groups all over the world which is mainly because of several ecological problems including deforestation and socio-economic problems related to tribal or native people associated with them.

The *Silent Valley* hydroelectric project was one of the first such projects situated in the tropical rain forest area of Western Ghats which attracted much concern of the people. *The crusade against the ecological damage and deforestation caused due to Tehri dam was led by Sh. Sunder Lal Bahuguna, the leader of Chipko movement. The cause of Sardar Sarovar Dam related issues has been taken up by the environmental activists Medha Patekar, joined by Arundhati Roy and Baba Amte.*

For building big dams, large scale devastation of forests takes place which breaks the natural ecological balance of the region. Floods, droughts and landslides become more prevalent in such areas. Forests are the repositories of invaluable gifts of nature in the form of biodiversity and by destroying them (particularly, the tropical rain forests) we are going to lose these species even before knowing them. These species could be having marvelous economic or medicinal value and deforestation results in loss of this storehouse of species which have evolved over millions of years in a single stroke.

Sources of Water

Water is an indispensable natural resource on this earth on which all life depends. About 97% of the earth's surface is covered by water and most of the animals and plants have 60-65% water in their body.

Water is characterised by certain unique features which make it a marvellous resource:

(i) It exists as a liquid over a wide range of temperature, i.e. from 0° to 100°C.

(ii) It has the highest specific heat, due to which it warms up and cools down very slowly without causing shocks of temperature jerks to the aquatic life.

(iii) It has a high latent heat of vaporisation. Hence, it takes a huge amount of energy for getting vaporised. That's why it produces a cooling effect as it evaporates.

(iv) It is an excellent solvent for several nutrients. Thus, it can serve as a very good carrier of nutrients, including oxygen, which are essential for life. But, it can also easily dissolve various pollutants and become a carrier of pathogenic microorganisms.

(v) Due to high surface tension and cohesion it can easily rise through great heights through the trunk even in the tallest of the trees like Sequoia.

(vi) It has an anomalous expansion behaviour, i.e. as it freezes, it expands instead of contracting and thus becomes lighter. It is because of this property that even in extreme cold, the lakes freeze only on the surface. Being lighter the ice keeps floating, whereas the bottom waters remain at a higher temperature and therefore, can sustain aquatic organisms even in extreme cold.

The water we use keeps on cycling endlessly through the environment, which we call as Hydrological Cycle. We have enormous resources of water on the earth amounting to about 1404 million Km^3. The water from various moist surfaces evaporates and falls again on the earth in the form of rain or snow and passes through living organisms and ultimately returns to the oceans.

Every year about 1.4 inch thick layer of water evaporates from the oceans, more than 90% of which returns to the oceans through the hydrological cycle. Solar energy drives the water cycle by evaporating it from various water bodies, which subsequently return through rainfall or snow. Plants too play a very important role by absorbing the groundwater from the soil and releasing it into the atmosphere by the process of transpiration.

Global distribution of water resources is quite uneven depending upon several geographic factors. Tropical rain forest areas receive maximum rainfall while the major world deserts occur in zones of dry, descending air (20-40° N and S) and receive very little rainfall.

Use and Misuse of Water: Due to its unique properties water is of multiple uses for all living organisms. Water is absolutely essential for life. Most of the life processes take place in water contained in the body. Uptake of nutrients, their distribution in the body, regulation of temperature, and removal of wastes are all mediated through water.

Human beings depend on water for almost every developmental activity. Water is used for drinking, irrigation, transportation, washing and waste disposal for industries and used as a coolant for thermal power plants. Water shapes the earth's surface and regulates our climate.

Water Use by Humans is of Two Types:

Water Withdrawal: taking water from groundwater or surface water resource and water consumption: the water which is taken up but not returned for reuse. Globally, only about 60 per cent of the water withdrawn is consumed due to loss through evaporation.

With increasing human population and rapid development,

the world water withdrawal demands have increased many folds and a large proportion of the water withdrawn is polluted due to anthropogenic activities. On a global average 70 per cent of the water withdrawn is used for agriculture. In India, we use 93% of water in agricultural sector while in a country like Kuwait, which is water-poor, only 4% is used for watering the crops. About 25% of water on global average is used in industry, which again varies from a high of 70% in European countries to as low as 5% in less developed countries. Per capita use of water shows wide variations. In USA, an average family of 4 consumes more than 1000 M^3 of water per year, which is many times more than that in most developing countries.

Precious Means of Nature: Although water is very abundant on this earth, yet it is very precious. Out of the total water reserves of the world, about 97% is salty water (marine) and only 3% is fresh water. Even this small fraction of fresh water is not available to us as most of it is locked up in polar ice caps and just 0.003% is readily available to us in the form of groundwater and surface water.

Overuse of groundwater for drinking, irrigation and domestic purposes has resulted in rapid depletion of groundwater in various regions leading to lowering of water table and drying of wells. Pollution of many of the groundwater aquifers has made many of these wells unfit for consumption.

Rivers and streams have long been used for discharging the wastes. Most of the civilisations have grown and flourished on the banks of rivers, but unfortunately, growth in turn, has been responsible for pollution of the rivers.

As per the United Nations estimates (2002), at least 101 billion people do not even have access to safe drinking water and 2.4 billion do not have adequate sanitation facilities. Increasing population and expanding development would further increase the demands for wastes. It is estimated that by 2024, two-thirds of the world population would be suffering from acute water shortage.

Groundwater: About 9.86% of the total fresh water resources is in the form of groundwater and it is about 35-50 times that of surface water supplies.

Till some time back groundwater was considered to be very pure. However, of late, even groundwater aquifers have been found to be contaminated by leachates from sanitary landfills, etc.

A layer of sediment or rock that is highly permeable and contains water is called an aquifer. Layers of sand and gravel are good aquifers while clay and crystalline rocks (like granite) are not since they have low permeability. Aquifers may be of two types:

Unconfined aquifers which are overlaid by permeable earth materials and they are recharged by water seeping down from above in the form of rainfall and snow melt.

Confined aquifers which are sand witched between two impermeable layers of rock or sediments and are recharged only in those areas where the aquifer intersects the land surface. Sometimes the recharged area is hundreds of kilometers away from the location of the well. Groundwater is not static, it moves, though at a very slow rate of about a meter or so in a year.

Influences of Groundwater

(i) *Subsidence:* When groundwater withdrawal is more than its recharge rate, the sediments in the aquifer get compacted, a phenomenon known as *ground subsidence.* Huge economic losses may occur due to this phenomenon because it results in the sinking of overlying land surface. The common problems associated with it include structural damage in buildings, fracture in pipes, reversing the flow of sewers and canals and tidal flooding.

(ii) *Lowering of Water Table:* Mining of groundwater is done extensively in arid and semi-arid regions for irrigating crop fields. However, it is not advisable to do excessive mining as it would cause a sharp decline in future agricultural production, due to lowering of water table.

(iii) *Water Logging:* When excessive irrigation is done with brackish water it raises the water table gradually leading to water-logging and salinity problems.

Surface Water: The water coming through precipitation (rainfall, snow) when does not percolate down into the ground or does not return to the atmosphere as evaporation or transpiration loss, assumes the form of streams, lakes, ponds, wetlands or artificial reservoirs known as surface water. The surface water is largely used for irrigation, industrial use, public water supply, navigation, etc. A country's economy is largely dependent upon its rivers.

Countries with Rich and Poor Water Resources: The top ten water rich countries are Iceland, Surinam, Guyana, Papua New Guinea, Gabon, Solomon Islands, Canada, Norway, Panama, and Brazil lying in the far north and have low evaporation losses.

The water poor countries include Kuwait, Egypt, United Arab Emirates, Malta, Jordon, Saudi Arabia, Singapore, Maldovia, Israel and Oman, lying in the desert belt at about 15° to 25° Latitude and some of them like Malta and Singapore are densely populated areas resulting in low per capita water.

Floods: In some countries like India and Bangladesh rainfall does not occur throughout the year, rather, 90% of it is concentrated into a few months (June-September). Heavy rainfall often causes floods in the low-lying coastal areas. Prolonged downpour can also cause the over-flowing of lakes and rivers resulting into floods. Deforestation, overgrazing, mining, rapid industrialisation, global warming, etc. have also contributed largely to a sharp rise in the incidence of floods, which otherwise is a natural disaster.

Floods have been regular features of some parts of India

and Bangladesh causing huge economic loss as well as loss of life. People of Bangladesh are accustomed to moderate flooding during monsoon and they utilise the flood water for raising paddy. But, severe floods like that in 1970, 1988 and 1991 resulting from excessive Himalayan runoff and storms, had very disastrous consequences causing massive deaths and damages. In 1970, about one million people were drowned while 1,40,000 people died in 1991. Networking of rivers is being proposed at national level to deal with the problems of floods.

Droughts: There are about 80 countries in the world, lying in the arid and semi-arid regions that experience frequent spells of droughts, very often extending up to year long duration. When annual rainfall is below normal and less than evaporation, drought conditions are created.

Ironically, these drought-hit areas are often having a high population growth which leads to poor land use and makes the situation worse.

Anthropogenic Causes: Drought is a meteorological phenomenon, but due to several anthropogenic causes like over grazing, deforestation, mining, etc. there is spreading of the deserts tending to convert more areas to drought affected areas. In the last twenty years, India has experienced more and more desertification, thereby increasing the vulnerability of larger parts of the country to droughts.

Erroneous and intensive cropping pattern and increased exploitation of scarce water resources through well or canal irrigation to get high productivity has converted drought-prone areas into decertified ones. In Maharashtra there has been no recovery from drought for the last 30 years due to over-exploitation of water by sugarcane crop which has high water demands.

Remedial Measures: Indigenous knowledge in control of drought and desertification can be very useful for dealing with the problem. Carefully selected mixed cropping help optimise production and minimise the risks of crop failures. Social Forestry and Wasteland development can prove quite effective to fight the problem, but it should be based on proper understanding of ecological requirements and natural process, otherwise it may even boomerang. The Kolar district of Karnataka is one of the leaders in Social Forestry with World Bank Aid, but all its 11 talukas suffer from drought. It is because the tree used for plantation here was *Eucalyptus* which is now known to lower the water table because of its very high transpiration rate.

Conflicts of Water Sharing: Indispensability of water and its unequal distribution has often led to inter-state or international disputes. Issues related to sharing of river water have been largely affecting our farmers and also shaking our governments. Some major water conflicts are discussed here.

Water-dispute in the Middle East: Three river basins, namely the Jordan, the Tigris-Euphrates and the Nile are the shared water resources for Middle East countries. Ethiopia controls the head waters of 80% of Nile's flow and plans to increase it.

Sudan too is trying to divert more water. This would badly affect Egypt, which is a desert, except for a thin strip of irrigated cropland along the river Nile and its delta. The population of Egypt is likely to double in the next 20 years, thereby increasing its water crisis. Likewise there is a fierce battle for water among Jordan, Syria and Israel for the Jordan River water share.

Turkey has abundant water and plans to build 22 dams on Tigris-Euphrates for Hydroelectric power generation. But, it would drastically reduce the flow of water to Syria and Iraq, lying downstream. Turkey dreams to become the region's water Super power. It plans to transport and sell water to starved Saudi Arabia, Kuwait, Syria, Israel and Jordan. Probably, the next war in the Middle East would be fought over water and not oil.

The Indus Water Treaty: The Indus, one of the mightiest rivers is dying a slow death due to dams and barrages that have been built higher upon the river. The Sukkur barrage (1932), Ghulam Muhammad Barrage at Kotri (1958) and Tarbela and Chasma Dams on Jhelum, a tributary of Indus have resulted in severe shrinking of the Indus delta. In 1960, the Indus water treaty was established by which Indus, the Jhelum and the Chenab were allocated to Pakistan and the Satluj, the Ravi and the Beas were allocated to India. Being the riparian state, India has pre-emptive right to construct barrages across all these rivers in Indian territory. However, the treaty requires that the three rivers allocated to Pakistan may be used for *non-consumptive* purposes by India, i.e. without changing its flow and quality. With improving political relations between the two countries it is desirable to work out techno-economic details and go for an integrated development of the river basin in a sustainable manner.

The Cauvery Water Dispute: Out of India's 18 major rivers, 17 are shared between different states. In all these cases, there are intense conflicts over these resources which hardly seem to resolve. The Cauvery river water is a bone of contention between Tamil Nadu and Karnataka and the fighting is almost hundred years old. Tamil Nadu, occupying the downstream region of the river wants water-use regulated in the upstream. Whereas, the upstream state Karnataka refuses to do so and claims its primacy over the river as upstream user. The river water is almost fully utilised and both the states have increasing demands for agriculture and industry.

The consumption is more in Tamil Nadu than Karnataka where the catchment area is more rocky. On June 2,1990, the Cauvery Water Dispute Tribunal was set up which through an interim award directed Karnataka to ensure that 205 TMCF of water was made available in Tamil Nadu's Mettur dam every year, till a settlement was reached. In 1991-92 due to good monsoon, there was no dispute due to good stock of water in Mettur, but in 1995, the situation turned into a crisis due to delayed rains and an expert Committee was set up to look into the matter which found that there was a complex cropping pattern in Cauvery basin.

Sambra paddy in winter, *Kurvai* paddy in summer and some cash crops demanded intensive water, thus aggravating the water crisis. Proper selection of crop varieties, optimum use of water, better rationing, rational sharing patterns, and pricing of water are suggested as some measures to solve the problem.

The Canal Dispute of Satluj-Yamuna Link (SYL): The issue of sharing the Ravi-Beas waters and SYL issue between Punjab and Haryana is being discussed in the Supreme Court. The Eradi Tribunal (1985) based the allocation of water on the basis of the time-inflow data of 20 years (1960-80), according to which 17.17 MAF (million acre feet) water was available. However, now it is argued by Punjab that in the last 17 years there has been consistent decline reducing the quantity to 14.34 MAF. The Supreme Court on January 15, 2002 directed Punjab to complete and commission the SYL within a year, failing which the Centre was told to complete it. However, two years have passed, but neither the SYL has been completed nor the conflict over sharing of Ravi-Beas water is resolved.

The conflict is that Punjab being the riparian state for Beas, Ravi and Satluj stakes its claim, Haryana has faced acute shortage of water after it became a state in 1966 and has been trying to help it out by signing an MOU (Memorandum of understanding) with UP, Rajasthan and Delhi for allocation of Yamuna waters. The Yamuna basin covers the state of Haryana while the Indus basin covers Punjab. The conflict revolving around sharing of river water needs to be tackled with greater understanding and objectivity.

System of Traditional Water Management: In India, even today, there are several villages where water management is done not by the Irrigation Department, but by local managers. In south India, a *neerkatti* manages the traditional tanks very efficiently based on his/her knowledge of the terrain, drainage and irrigation needs. They usually give preference to the tail end fields and decide per capita allocation of water based on the stock of available water in the tank and crop needs. In Maharashtra, the water mangers are called *havaldars* or *jaghyas* who manage and resolve conflicts by overseeing the water channels from main canal to the distributary canals. In Laddakh, the water manager is known as *churpun* who has got complete charge with full powers over allocation of available water.

The major source of water is melt water from glaciers and snow supplementary by water from springs and marshes. The water is distributed to different fields through an intricate network of earthen channels. In traditional water management, innovative arrangements ensure equitable distribution of water, which are democratically implemented. The 'gram-sabhas' approve these plans publicly. While water disputes between states and nations often assume battle like situations, our traditional water managers in villages prove to be quite effective.

Uses and Harms of Big Dams

Benefits: River valley projects with big dams have usually been considered to play a key role in the development process due to their multiple uses. India has the distinction of having the largest number of river-valley projects. These dams are often regarded as a symbol of national development. The tribals living in the area pin big hopes on these projects as they aim at providing employment and raising the standard and quality of life. The dams have tremendous potential for economic upliftment and growth. They can help in checking floods and famines, generate electricity and reduce water and power shortage, provide irrigation water to lower areas, provide drinking water in remote areas and promote navigation, fishery, etc.

Problems of Atmosphere: The environmental impacts of big-dams are also too many due to which very often the big dams become a subject of controversy. The impacts can be at the upstream as well as downstream levels.

The upstream problems include the following:

(i) Displacement of tribal people
(ii) Loss of forests, flora and fauna
(iii) Changes in fisheries and the spawning grounds
(iv) Siltation and sedimentation of reservoirs
(v) Loss of non-forest land
(vi) Stagnation and waterlogging near reservoir
(vii) Breeding of vectors and spread of vector-borne diseases
(viii) Reservoir induced seismicity (RIS) causing earthquakes
(ix) Growth of aquatic weeds.
(x) Microclimatic changes.

The downstream impacts include the following:

(i) Water logging and salinity due to over irrigation
(ii) Micro-climatic changes
(iii) Reduced water flow and silt deposition in river
(iv) Flash floods
(v) Salt water intrusion at river mouth
(vi) Loss of land fertility along the river since the sediments carrying nutrients get deposited in the reservoir.
(vii) Outbreak of vector-borne diseases like malaria. Thus, although dams are built to serve the society with multiple uses, but it has several serious side-effects. That is why now there is a shift towards construction of small dams or mini-hydel projects.

Minerals and their Importance

Minerals are naturally occurring, inorganic, crystalline solids having a definite chemical composition and characteristic physical properties. There are thousands of minerals occurring in different parts of the world. However, most of the rocks, we see everyday are just composed of a few common minerals like quartz, feldspar, biotite, dolomite, calcite, laterite, etc. These minerals, in turn, are composed of some elements like silicon, oxygen, iron, magnesium, calcium, aluminium, etc.

Uses and Exploitation: Minerals find use in a large number of ways in everyday use in domestic, agricultural, industrial and commercial sectors and thus form a very important part of any nation's economy. The main uses of minerals are as follows:

(i) Development of industrial plants and machinery.

(ii) Generation of energy e.g. coal, lignite, uranium.

(iii) Construction, housing, settlements.

(iv) Defence equipments-weapons, armaments.

(v) Transportation means.

(vi) Communication-telephone wires, cables, electronic devices.

(vii) Medicinal system-particularly in Ayurvedic System.

(viii) Formation of alloys for various purposes (e.g. phosphorite).

(ix) Agriculture—as fertilizers, seed dressings and fungicides (e.g. zineb containing zinc, Maneb-containing manganese, etc.).

(x) Jewellery—e.g. Gold, silver, platinum, diamond.

Based on their properties, minerals are basically of two types:

(i) Non metallic minerals e.g. graphite, diamond, quartz, feldspar.

(ii) Metallic minerals e.g. Bauxite, laterite, haematite, etc.

Use of metals by human beings has been so extensive since the very beginning of human civilisation that two of the major epochs of human history are named after them as Bronze Age and Iron Age. The reserves of metals and the technical know-how to extract them have been the key elements in determining the economy and political power of nations. Out of the various metals, the one used in maximum quantity is Iron and steel

(740 million metric tons annually) followed by manganese, copper, chromium, aluminium and Nickel.

Distribution and uses of some of the major metallic and non-metallic minerals are given in Tables.

Major Reserves and Important uses of Some of the Major Metals

Metal	*Major World Reserves*	*Major Uses*
Aluminium	Australia, Guinea, Jamaica	Packaging food items, transportation, utensils, electronics
Chromium	CIS, South Africa	For making high strength steel alloys, In textile/tanning industries
Copper	U.S.A., Canada, CIS, Chile, Zambia	Electric and electronic goods, building, construction, vessels
Iron	CIS, South America, Canada, U.S.A.	Heavy machinery, steel production transportation means
Lead	North America, U.S.A., CIS	Leaded gasoline, Car batteries, paints, ammunition
Manganese	South Africa, CIS, Brazil, Gabon	For making high strength, heat-resistant steel alloys
Platinum group	South Africa, CIS	Use in automobiles, catalytic converters, electronics, medical uses.
Gold	South Africa, CIS, Canada	Ornaments, medical use, electronic use, use in aerospace
Silver	Canada, South Africa, Mexico	Photography, electronics jewellery
Nickel	CIS, Canada, New Caledonia	Chemical industry, steel alloys

Major uses of Some Non-metallic Minerals

Non-metal Mineral	***Major Uses***
Silicate minerals	Sand and gravel for construction, bricks, paving, etc.
Limestone	Used for concrete, building stone, used in agriculture for neutralizing acid soils, used in cement industry
Gypsum	Used in plaster wall-board, in agriculture
Potash, phosphorite	Used as fertilizers
Sulphur pyrites	Used in medicine, car battery, industry.

It is evident from the Tables that the CIS countries

(The Commonwealth of Independent States, *i.e.* 12 republics of former USSR), the United States of America, Canada, South Africa and Australia are having the major world reserves of most of the metallic minerals.

Due to huge mineral and energy resources, the USA became the richest and the most powerful nation in the world in even less than 200 years. Japan too needs a mention here, as there are virtually no metal reserves, coal, oil and timber resources in Japan and it is totally dependent on other countries for its resources. But, it has developed energy efficient technologies to upgrade these resources to high quality finished products to sustain its economy.

Minerals are sometimes classified as Critical and Strategic.

Critical minerals are essential for the economy of a nation e.g. iron, aluminium, copper, gold, etc.

Strategic minerals are those required for the defence of a country e.g. Manganese, cobalt, platinum, chromium, etc.

Few Principal Minerals of India

(a) Energy generating minerals:

Coal and Lignite: West Bengal, Jharkhand, Orissa, M.P., A.P. *Uranium* (Pitchblende or Uranite ore): Jharkhand, Andhra Pradesh (Nellore, Nalgonda), Meghalaya, Rajasthan (Ajmer).

(b) Other commercially used minerals:

Aluminium (Bauxite Ore): Jharkhand, West Bengal, Maharashtra, M.P., Tamil Nadu.

Iron (Haematite and Magnetite Ore): Jharkhand, Orissa, M.P., A.P., Tamil Nadu, Karnataka, Maharashtra and Goa.

Copper (Copper Pyrites): Rajasthan (Khetri), Bihar, Jharkhand, Karnataka, M.P., West Bengal, Andhra Pradesh and Uttaranchal.

Impacts of Mineral Extraction and Use on Environment : The issue related to the limits of the mineral resources in our earth's crust or in the ocean is not so significant. More important environmental concern arises from the impacts of extraction and processing of these minerals during mining, smelting, etc.

Indian Scenario: India is the producer of 84 minerals the annual value of which is about Rs. 50,000 crore. At least six major mines need a mention here which are known for causing severe problems:

(i) Jaduguda Uranium Mine, Jharkhand—exposing local people to radioactive hazards.

(ii) Jharia coal mines, Jharkhand—underground fire leading to land subsidence and forced displacement of people.

(iii) Sukinda chromite mines, Orissa—seeping of hexavalent chromium into river posing serious health hazard, Cr^{6+} being highly toxic and carcinogenic.

(iv) Kudremukh iron ore mine, Karnataka—causing river pollution and threat to biodiversity.

(v) East coast Bauxite mine, Orissa—Land encroachment and issue of rehabilitation unsettled.

(vi) North-Eastern Coal Fields, Assam—Very high sulphur contamination of groundwater.

Impacts of Mining: Mining is done to extract minerals (or fossil fuels) from deep deposits in soil by using sub-surface mining or from shallow deposits by surface mining. The former method is more destructive, dangerous and expensive including risks of occupational hazards and accidents.

Surface mining can make use of any of the following three types:

(a) *Open-pit Mining:* in which machines dig holes and remove the ores (e.g. copper, iron, gravel, limestone, sandstone, marble, granite).

(b) *Dredging:* in which chained buckets and draglines are used which scrap up the minerals from under-water mineral deposits.

(c) *Strip Mining:* in which the ore is stripped off by using bulldozers, power shovels and stripping wheels (e.g. phosphate rocks).

The environmental damage caused by mining activities are as follows:

(i) *De-vegetation and De-facing of Landscape:* The topsoil as well as the vegetation are removed from the mining area to get access to the deposit. While large scale deforestation or de-vegetation leads to several ecological losses as already discussed in the previous section, the landscape also gets badly affected. The huge quantities of debris and tailings along with big scars and disruptions spoil the aesthetic value of the region and make it prone to soil erosion.

(ii) *Subsidence of Land:* This is mainly associated with underground mining. Subsidence of mining areas often results in tilting of buildings, cracks in houses, buckling of roads, bending of rail tracks and leaking of gas from cracked pipelines leading to serious disasters.

(iii) *Groundwater Contamination:* Mining disturbs the natural hydrological processes and also pollutes the groundwater. Sulphur, usually present as an impurity in many ores is known to get converted into sulphuric acid through microbial action, thereby making the water acidic. Some heavy metals also get leached into the groundwater and contaminate it posing health hazards.

(iv) *Surface Water Pollution:* The acid mine drainage often contaminates the nearby streams and lakes. The acidic water is detrimental to many forms of aquatic life. Sometimes radioactive substances like uranium also contaminate the water bodies through mine wastes and kill aquatic animals. Heavy metal pollution of water bodies near the mining areas is a common feature creating health hazards.

(v) *Air Pollution:* In order to separate and purify the metal from other impurities in the ore, smelting is done which emits enormous quantities of air pollutants damaging the vegetation nearby and has serious environmental health impacts. The suspended particulate matter (SPM), SOx, soot, arsenic particles, cadmium, lead, etc. shoot up in the atmosphere near the smelters and the public suffers from several health problems.

(vi) *Occupational Health Hazards:* Most of the miners suffer from various respiratory and skin diseases due to constant exposure to the suspended particulate matter and toxic substances. Miners working in different types of mines suffer from asbestosis, silicosis, black lung disease, etc.

Remedial Measures: Safety of mine workers is usually not a priority subject of industry. Statistical data show that, on an average, there are 30 non-fatal but disabling accidents per ton of mineral produced and one death per 2.5 tons of mineral produced. In order to minimise the adverse impacts of mining it is desirable to adopt eco-friendly mining technology. The low-grade ores can be better utilised by using microbial-leaching technique. The bacterium *Thiobacillus ferroxidans* has been successfully and economically used for extracting gold embedded in iron sulphide ore. The ores are inoculated with the desired strains of bacteria, which remove the impurities (like sulphur) and leave the pure mineral. This biological method is helpful from economic as well as environmental point of view. Restoration of mined areas by re-vegetating them with appropriate plant species, stabilisation of the mined lands, gradual restoration of flora, prevention of toxic drainage discharge and conforming to the standards of air emissions are essential for minimizing environmental impacts of mining.

Trees of Special Care

It involves the improvement of our forests to ensure regular supply of useful plant products. There is requirement for improving many forest plants genetically for better yield, quick growth, resistance to pests and pathogens. It involves the following measures.

(i) *Improvement Cutting:* It involves cutting of all old dying trees, trees damaged by storms and animals, diseased trees and crooked trees.

(ii) *Selective Cutting:* It is the cutting of timber trees when they are fully matured. It is better to remove the less vigorous trees for timber and allow the others a chance for normal growth.

(iii) Forest floor should be quickly cleared off from fallen tree and broken branches.

(iv) Overcrowding of trees in the forest should be avoided.

(v) Afforestation Programmes.

Many afforestation programmes are undertaken by our government which are of three types.

Social Forestry Programme (1976): It was started in India in 1976 by NCA (National Commission on Agriculture). It involves the afforestation on public and common land for fuel wood, fodder, agricultural implements and fruits for the benefit of rural community.

The programme is aimed at reducing the demand on existing forests. The plants selected for social forestry vary from place to place. The important ones are Albizia, Moringa, Delbergia, Acacia, Casuarina, Tectona, Zizyphus, Morus, etc.

Agro-forestry Programme: It is growing of multipurpose trees, shrubs, horticultural plants and forage plants alongwith crops. It is aimed at reviving the ancient practice of using the same land for fanning, forestry and animal husbandry.

Urban Forestry Programme: It is aimed at growing of fruit, flower and shade bearing shrubs and trees in urban areas along roads, in private compounds and in common parks. Forest conservation Act, 1980 was enforced to check indiscriminate deforestation.

Great Enterprises

Just as man has important organ for breathing same is the case with earth. To save important part of the mother earth some people became conscious and started opposing the selfish people who not only destroyed environment of the earth but also destroyed themselves as well as other organisms. Chipko Andolan is similar movement in which people opposed the cruelties of man on the forest or nature.

This Chipko Andolan started in the year 1972 in the upper reaches of Uttar Pradesh Himalayan region, Tehri Garhwal. Alaknanda valley faced unprecedented flood due to degradation of forests on the hills by the contractors.

The menfolk opposed this large scale deforestation but never got success. The real movement started in March 1973 in Gopeshwar in Chamoli district when a sports good factory of Allahabad was to cut ten Ash trees near the village Mandal. The local people prevented the same by hugging the marked trees (hence chipko).

The same was repeated later in Rampur Phata in 1974, women led by Gaura Devi successfully prevented felling of forest trees near village Reni. The movement spread slowly to all nearly areas and outside by Chandi Prasad Bhatt of Gopeshwar and Sunder Lal Bahuguna of Silyara in Tehri region. Sunder Lal Bahuguna who is now a world famous figure on eco-protection conducted a peace march of about 3000 km in this hilly region to protest against the felling of trees and to prevent the construction of hydroelectric project in the silent valley region in Kerala. Such movements will help to preserve forests for future India. The Chipko movement succeeded and following are the results of this movement.

(i) Contractor system abolished by the Government.

(ii) Uttar Pradesh Forest Development Corporation department formed which works for the welfare of the hilly areas and the population residing there.

(iii) It slows down the process of deforestation.

(iv) It created awareness amongst the people of not only U.P. but of other states.

(v) It enlightened people about the necessity of ecological balance in nature.

Resources from Flora

Forests are self-sustained wooden tracts with biotic community dominated by trees forming a sort by canopy. India gave birth to Gautam Buddha, in a sacred grove of "SAL Tress" dedicated to the Goddess "LUMBINI." He achieved salvation under a pipal tree forest in a peculiar organism of unlimited kindness and benevolences, that makes no demands for its sustenance and extend generosity through its products of life activity.

Areas in Use

India has a very rich Flora and Fauna much of which is present in forest area. The flora and Fauna of a forest are inter dependent on each other. The plant wealth of India's forests are made up of about 45000 species of trees, shrubs, herbs, creepers and climbers which account for about 12% of global plant wealth. Indian forest contain 7500 species of animals and as many as 60,000 species of insect pests.

The forests are complete ecosystem consisting of two components abiotic and biotic, i.e. both non-living and living components.

The Structure

The non-living factors prevailing in an ecosystem from abiotic components.

1. *Water:* It is a complex mixture of many compounds and gases. Mainly it is formed of two molecules hydrogen

and oxygen. It forms 71% of the earth's surface and represents the most extensive habitat for the organisms. It helps in growth of the plants.

2. *Soil:* It is uppermost crust of the earth which is rich in organic and inorganic matter. It sustains the life of the plants on the earth. Plants absorbs essential nutrients from the soil-

3. *Climate:* It plays important role as it influences the dense growth, high yield and stimulates the other micro organism to form the hymns. Suitable climate of the region effects the forest canopy and increases forest products. It includes.

   (i) *Light:* It provides solar energy to the ecosystem for heating and photosynthesis. Besides photosynthesis, light controls morphogenesis. Photoperiods influence leaf fall, appearance of new leaves and flowering in plants.

   (ii) *Temperature:* Suitable temperature stimulates the rich growth of plants and transportation rate in which the plants give out excess water into atmosphere in the form of vapours. Most organisms survive a temperature range of 0° to 60° C less.

   (iii) *Precipitation:* It is the process of condensation of water vapour which evaporate in the atmosphere through transpiration of forest trees. It helps in the formation of clouds, snow, hails, etc.

   (iv) *Humidity:* Transpiration is inversely related to humidity. Humid areas have abundant dense plant, growth with numerous animals and epiphytes. Humidity maintains the soil moisture which helps in improving the quality of forests.

   (v) *Wind:* It controls weather, transpiration, pollination and dissemination of propagules.

   (vi) *Atmospheric Gases:* These include oxygen, nitrogen

and carbondioxide. The carbonoxide is used in the process of photosynthesis. Nitrogen increases the fertility of the soil and oxygen is used for respiration thus, play important role in maintaining forest ecosystem.

The living organisms present in the forest form the biotic components. These include plants and animals. These are respectively called producers, consumers and decomposers.

(i) *Producers:* They are the plants in the form of grasses, herbs, shrubs and trees. These synthesise their own food by the process of photosynthesis. The inorganic material required are obtained from the forest ecosystem.

(ii) *Consumers:* These are mainly the animal which are unable to synthesise their food. Therefore, they take other organisms or parts of organisms. These are further divided into:

(a) *Herbivores (Primary Consumers):* These are the animals which feed on the green plants only e.g. deer, goat, rabbit.

(b) *Carnivores (Secondary Consumers):* These are animals which take herbivores e.g. frogs, snake, wolf.

(c) *Top Carnivores (Tertiary Consumers):* These are largest carnivores which feed on small carnivores e.g. Lion.

(iii) *Decomposers:* These are mainly bacteria and fungi, which obtain their food from organic materials of dead producers and consumers.

The Role

Forests are invaluable wealth of a country and a renewable natural resource. They are beneficial to man in several ways:

1. Forests are source of wood which is used as timber. Timber is the basic material for building houses making furniture, railways sleepers, sport goods, boats, ships, etc.

2. The forest provide fuel in the form of wood for cooking and keeping warm. Over 1500 million people use wood as a source of energy. Over 1000 million m^3 of wood is used as fuel in the world. In S-E. Asia fuel wood provides 42% of the total energy requirements. It is 58% of the total in Africa.
3. The forest provide food to the tribes in the form of tubers, roots, leaves and fruits.
4. Forest provides a large number of important products such as essential oils, (Kas sandal wood), natural camphor (Laurel tree-Lauru nobilis), tannins, resins, gums, dyes drug, spices, bidi wrappers (tendi leaves), soap substitutes like ritha and sikakai, lac, honey, wax, silk, etc. Horns hides, musk and ivory are provided by forest animals.
5. Bamboos form the poor's man timber. They are used in carts ropes, roofing, flooring, scaffolding, etc.
6. Forests keep the environment cool through increasing humidity of the atmosphere by transpiration which increases the chances of rainfall.
7. The fertility of the forest soil increases by dead fallen leaves which decay and form hymns.
8. Forests maintain oxygen contents of the atmosphere.
9. Forests reduce atmospheric pollution by using carbon dioxide in the process of photosynthesis.
10. Forest trees provide shelter to a number of animals such as insects reptiles, birds, mammal.
11. Forest prevent soil erosion by:
    (i) The roots which firmly binds the soil,
    (ii) Their fallen leaves which decrease the speed of water flow,
    (iii) Their trees which decrease the velocity of wind to prevent dislodging of the soil particle.

12. Forests have aesthetic importance as they simulate art of literature and music.
13. The humus of the forests floor acts as a giant sponge which absorbs large amount of water. This regulates the water flow in streams, springs, wells and rivers.

Botanical World

According to Brewbaker (1984) total forest area of the world in 1900 was nearly 1000 million hectares (mha). By 1975 it was reduced to 2890 Mha and if this continues like this by 2000 A.D. it would be merely 2370 Mha. In India, nearby 80% of the country was covered with forests around 3000 B.C. Waves of migrants came to India, many from treeless countries and changed the whole landscape.

There are no reliable statistics of forest cover in India. However according to Central Forestry Commission (1980), there are only 74.8 Mha of the land area under forest cover which forms about 22.7% of the total land mass of India. It has increased to 75.23 Mha (22.96%) in 1993.

In India there are 16 types of forests on the basis of geographical, climatic and edaphic conditions and their species composition. The most common in the tropical dry deciduous (38.7%) followed by tropical moist deciduous (30.9%) type. These two types of tropical deciduous forests contribute to the bulk (69.6) of the forest area in India.

As indicated earlier, according to CFC (1980) the forest cover in nearly 22.7%. However this is contrary to the general belief that effective forest cover is substantially low and that the forest wealth is deteriorating at rapid rate. Maximum damage in suffered by closed forests, which are more valuable from ecological and economic view points. We cannot afford depletion of these forests. The rapid decline of forests were:

(a) Overpopulation

(b) Deforestation

The Conservation

Deforestation is deterioration, reduction or removal of forest cover. It is the main cause of soil erosion in India. It is estimated that forest cover has declined from 7000 Mha in 1900 to 2890 Mha in 1985. It is expected that by 2000 A.D. forest cover will left only 2370 Mha. The reason for deforestation are given below:

Causes of Deforestation

1. *Population Explosion:* Population growth poses serious threat to the forest. The forests are the basis needs of everyday life as they provide us food, shelter and raw material for other essentialities but these forests deforested for fulfilling the increasing demands of over population like clearing of forests for agriculture, industries and urban area, etc.
2. *Fires:* Fire is worst enemy of the forests. Forests fire destroys adult as well as young trees and even seeds and useful fauna of the forests. Animal like is also lost and there is a danger to human life too in a forest fire.
3. *Pests:* Certain pests cause much damage in the forest. These either eat their leaves or bore into their stem or spread diseases.
4. *Industrial Development:* Industrial development although increases the economy of the country, but the basic raw material for these are obtained by the forests and ultimately growing industries for fulfilling their requirements started over exploiting the forests.
5. *Grazing and Growing Mammals:* These animals trample the seedlings or young plants, then destroy the leaves on the lower branches of tall trees and finally damage their trunks and roots.
6. *Agriculture Expansion:* Agriculture expansion means to convert the forest land area which are more fertile into agricultural areas to fulfil the crop necessities of the

people of the country. Shifting cultivation also leads to deforestation in which a patch of forest land is cleared through lopping and burning the remainder. The ash is mixed in the soil. The cleared land is cultivated for one year and next year that land is left bared and another forest area cleared. This type of cultivation is called Jhum in north-eastern region, Podu in Andhra Pradesh bewar or dahza in M.P. Every year about 5 lakh hectares of land are cleared for shifting cultivation.

7. *Hydroelectric Projects:* Dams, reservoirs and hydroelectric projects submerge large forest tracts, killing all plants and animals of the area.
8. *Quarrying and Mining:* In forest and hilly areas mining and quarrying are extremely harmful as they spoil vegetation over large areas.
9. *Short Term Policies:* Short term policies made by our Government, i.e. five year plans also adversely affect the forest areas. It is easy to cut the forest areas in five years but not easy to grow in such a short tenure. Thus these policies play significant role in deforestation.

Effects of Deforestation

(i) Increased chances of soil erosion.

(ii) Floods and draughts become more frequent.

(iii) Man deprived of benefits of trees and woodland animals.

(iv) Increased chances of reduced rainfall but more dust storms.

(v) Extinction of forest dwelling species.

(vi) Economic life of forest dwelling people is adversely affected.

(vii) Pattern of rain fall is changing.

(viii) Scarcity of timber, fuel wood, etc. Shortage of fuel wood forces people burn animal dung which otherwise have been used as manure. This reduces soil fertility.

SECURED FLORA

It is the development and management of forests in such a way to improve climate and water flow, prevent soil erosion and to provide optimum sustainable yield both for the present and future generations. The following steps can help conserve the forest wealth.

(i) Economically utilisation of timber to prevent wastage.

(ii) The wood fuel should be supplemented with alternative source of energy such as biogas.

(iii) Afforestation and reforestation should be undertaken. It is the development of forest where they did not exist in past (afforestation) or where they had been destroyed previously (reforestation). Afforestation should be done in areas unfit for agriculture, along highways and rivers, around playgrounds and parks and in spacious grounds. A special function of tree plantation, Van Mahotsava, is held each year in our country. In afforestation, native species should be preferred. Exotic species should be grown only on selective basis. NAEB (National Afforestation and Eco-development Board) looks after afforestation, tree plantation, restoration and development of ecologically fragile area.

(iv) Forest fires should be controlled through modern fire-fighting equipment.

(v) Using chemical and biological methods pests and diseases of forest trees should be controlled.

(vi) Grazing should be regulated and discouraged.

(vii) States forest policy should be adopted within overall frame work of National Forest Policy of 1988 which states that one third of total geographical area of country should be under forest cover.

(viii) Financial assistance to forestry sector should be increased as it was only 1.03% of total plan out lay in 7th five year plan and was further reduced to 0.94% in 8th five year plan.

6

EARTH AND ECOLOGY

A. How Universe came into being?

"Big bang", 8×10^9 years ago, estimated from galaxy expansion and background temperature of microwaves. Recent hypothesis suggests "bubble" inflating into different dimensions from a discontinuity ("white hole"-a tiny region of ultra high density and temperature) in a previous universe.

B. How Solar System came into being?

Aggregation of cold gas and dust by gravity 4-5 $\times 10^9$ years ago, estimated by radioactive decay (ratios of radioactive elements to their decomposition products).

Original gas (stellar): H_2, He, NH_3, H_2O, CH_4

Spinning of matter caused gravitational attraction—condensation of matter, increases spinning and gravitational attraction thereby attracting more matter—eventually, sun and planets formed via accretion of materials. Small planets lose H_2, He; gravitation and radioactive decay heats their cores, expelling N_2 (NH_3), H_2O, CO_2, CO, (CH_4) into the atmosphere; also H_2S, H_3PO_4 and rocks.

Inner Planets—heavier elements with high melting and boiling temperatures Earth-Composition- MgO, SiO_2, FeO, Fe, Ni, some water, NH_4

Outer Planets - lighter elements with low melting and boiling temperatures. Jupiter, Saturn, Uranus, Neptune - mainly composed of hydrogen, methane, ammonia ices Earth and Oceans formed 4.5 billion years ago

- decay of radioactive elements generates lots of heat
- gravity pulled most of the iron inward to form the planets core
- lighter elements, silicon, magnesium, aluminium, and oxygen bonded to form less dense compounds that migrated toward the surface
- by 4.0 billion years ago earth's surface cooled enough for water to exist primarily as a liquid. Atmosphere formed - formed 4.5 to 3.5 billion years ago - primary constituents of the atmosphere: methane, ammonia, carbon dioxide, water vapor. Carbon dioxide combines with water to form carbonic acid $H_2O + CO_2$ H_2CO_3 which dissolves rocks releasing minerals (nutrients), or the essential building blocks needed for life.

C. Organic Molecules Creation

Simple chemical reactions, requiring only raw materials and energy were the first step in the origin of life on earth.

Origin of Life: It took place 3.5 billion years ago.

Definition of life - capable of growing more complex, obtains and use energy by breaking down chemical compounds, capable of self replicating. Water is an ideal medium for biochemical reactions, its unique properties make it a prerequisite for origin of life, some of them are:

1. It retains heat
2. Very good solvent - dissolves many chemicals
3. Very good transport medium for nutrients

Primordial soup was created from a mixture of water vapor, ammonia, methane, and hydrogen.

Possible sources of energy for the chemical reactions:

1. Electrical discharge
2. Ultraviolet light
3. Geothermal and hydrothermal heat (ocean floor)

D. How Self-producing System came into being?

Amino acids, nucleotides form polymers: some theoreticians suggest that life must have both self-reproducing templates (nucleic acids) and catalysts (proteins) -which came first? How did the two get together?

Note that some RNAs (viral) are self-templates and some show enzymatic activity (RNA as selfsplicing). Because of this, some speculate that RNA was the first self-reproducing system. Some amino acids, trinucleotides (model colons) show affinities, suggesting a natural system leading to nucleic acid-directed protein synthesis. DNA evolved as a more stable form of template.

E. Gradual Transformation of Energy Metabolism

Strategy: use easiest energy sources first.

1. Fermentation (*e.g.*, anaerobic bacteria)

   Energy from UV, lightning stored as covalent bonds in soup components. First cells obtained energy by breaking down existing molecules (non-oxidative, because there was no free O_2). Eventually, cells would run out of substrate, since life grows faster than synthesis works.

2. Photosynthesis (*e.g.*, blue-green algae)

   Use visible light (not UV) as energy source (visible penetrates through H_2O farther than UV). Use H_2, H_2S and eventually H_2O as source of electrons and H+, depending on abundance. When HO used, O_2 produced;

production of O_2 leads to formation of O_3, shielding of UV opening of new niches in shallow water, land.

3. Aerobic respiration (*e.g.*, aerobic bacteria)

   Use energy stored in organic compounds. O_2 released by the oxidative reaction. Use is made possible by H_2O photosynthesis: possibly started in plants as protection against O_2. Efficient aerobic respiration allowed evolution of consumers.

F. Transformation of Complex Organelles (eucaryotes)

1. *Endosymbiotic Hypothesis:* Some organelles like plastids, mitochondria have double membranes, their own DNA, their own ribosomes, bacterial-like protein synthesis (similar but not identical). Suggest plastids formed by phagocytosis of photosynthetic procaryote, with stabilised symbiotic relationship (food, protection). Similarly mitochondria formed the same way with O_2-using procaryote.
2. *Packaging of Pre-existing Cell Components Hypothesis:* Photosynthetic cell with genes for chlorophyll, photosynthetic electron transport forms nucleus by invagination of plasma membrane; nucleus buds off some DNA to make accessory nucleus with thylakoids; mitochondria are evolved as modified chloroplasts.

Earth: For study purposes earth can be divide into following groups:

Geosphere: The geosphere is the solid Earth that includes continental and oceanic crust as well as the various layers of the Earth's interior. Ninety-four per cent of the earth is composed of the elements oxygen, iron, silica, and magnesium, and the interior of the earth is layered both chemically and mechanically. The geosphere is not static, but its surface (crust) is in a constant state of motion that gives rise to movement of the continents. The unifying theory that explains how continental drift is possible is called plate tectonics, it was developed by geologists in the 1960s.

Atmosphere: Each year volcanism spews significant amounts of gases (water vapor, carbon dioxide, sulphur dioxide, hydrogen, hydrogen sulphide, etc.) into the atmosphere. For example, explosive volcanism that injects large amounts of sulphur dioxide to the upper atmosphere is known to result in cooling of climate in the years immediate following the eruption.

Hydrosphere: Water is the main agent of chemical and mechanical erosion of the earth surface. When water flows over land it mechanically breaks rocks into finer particles and can chemically dissolve elements contained in rock-forming minerals.

The sediment and chemical constituents are carried away downstream as the water makes its way back to the ocean. For example, the awesome depth (in Kilometers) of the Grand Canyon which was cut by the erosive power of the Colorado River. Over long periods of geologic time, erosion by water can wear down mountain ranges (process called peneplanation) and transport this material to be deposited in sedimentary basins.

Biosphere: The weathering of the geosphere to form soils provides terrestrial plants with a firm substrate and vital nutrients (phosphorous, nitrogen) and minerals needed for plant growth. In addition, the chemical weathering of the geosphere by water transports essential nutrients (phosphorous, nitrogen, silica, etc.) to the oceans which are used by algae (marine plants) during photosynthesis.

Classification of Ecosystem

An ecosystem consists of all of the organisms living in a community plus all of the abiotic factors with which they interact. As such, ecosystems are the most important level of biological organisation.

Trophic Structure: Trophic structures determine the feeding relationships between organisms and the energy and chemical cycling pathways. Trophic structure is typically divided into several trophic levels. These levels are arranged hierarchically. The trophic level that supports all others are the primary producers of the ecosystem. These are the photosynthetic autotrophs that use sunlight to create complex carbohydrates.

All other organisms in an ecosystem are consumers, or heterotrophs, that depend directly or indirectly on the primary producers. Those that depend directly on primary producers are the organisms in the next trophic level, the primary consumers (herbivores). Animals that feed on herbivores (carnivores) occupy the next higher trophic level of the secondary consumers. Animals that feed on secondary consumers are tertiary consumers, and so on. Some consumers feed on the waste products or dead remains of members of other trophic levels. These are the decomposers (detritivores). The simplest arrangement of these trophic levels is: Primary producer - primary consumer - secondary consumer - tertiary consumer this is a form of a food chain. More often than not there are many organisms interacting at different trophic levels so interactions in the real world more often resemble a food web.

I. Energy in Ecosystems

1. *Global Energy Budget:* The primary source of energy for nearly all ecosystems on earth is the sun. Only a fraction of the sun's energy that strikes the earth reaches photosynthetic organisms and of this amount, about 1-2% is converted to chemical energy by photosynthesis. The amount of sunlight converted to chemical energy by primary producers during any given time is known as primary productivity.

   Different ecosystems have different productivities. Tropical rain forests and algal beds and reefs are extremely productive ecosystems.

2. *Limits to Productivity:* Seasonal limits like rainfall, temperature, and light intensity are very important.

Inorganic nutrients are also important. If a single factor is limiting it is known as the limiting nutrient. Replacing limiting nutrients in agricultural systems is at the heart of the notion of fertilizing.

3. *Energy Loss:* Energy is lost as it flows from one trophic level to the next. No consumer is ever 100% efficient in converting resources obtained from one trophic level and converting it into the energy needed to sustain life. There is only about a 10% transfer of energy from one trophic level to the next. For example, a caterpillar might consume 200 Joules of energy in plant material. Of that 200 J, 100 J might be eliminated as feces, 67 J as respiration, and just 33 J used for growth. It is that 33 J used for growth that will be available to an organism on another trophic level that eats the caterpillar. The differential between respiration and growth is even more extreme in homeotherms (like humans) that use a great deal of the energy they consume to maintain body temperature.

   Ecological efficiencies (the fraction of energy transferred from one trophic level to the next) can be thought of in terms of a pyramid of productivity, where the trophic levels are stacked in blocks. Pyramids like this are bottom heavy, with either much more biomass or many more numbers of organisms in the lower levels than the upper ones. This arrangement reflects the inherent inefficiencies that exist in transferring energy from one level to the next.

II. Biogeochemical Cycles

Life on earth receives the energy it needs from the sun which is, for all practical purposes, inexhaustible. The other requirements of life, in other words the essential chemical elements, are present in fixed amounts and so need to be constantly recycled. Chemicals that undergo biogeochemical cycles include water, nitrogen, phosphorus, carbon, etc.

A general model of biogeochemical cycling includes four parts divided between inorganic and organic materials and whether or not the chemical is available or unavailable for use by living things. These four parts, or compartments, are thus:

- Organic available: living things and detritus.
- Organic unavailable: typically fossilized living things (oil, peat, coal, etc.).
- Inorganic available: chemicals dissolved in water or present in the air that organisms can use.
- Inorganic unavailable: chemicals tied up in rocks.

Nutrient cycling is therefore the relative movement of matter between and among the various compartments.

A. *The Water Cycle:* All living things need water. Water also participates in all of the other chemical cycles. Most of the earth's water is present in the oceans. Most of it evaporates from and precipitates over these same oceans. Some evaporated water reaches land and falls as rain. Here some evaporates again. A larger fraction is turned into water vapor by plant transpiration. Another fraction of it percolates into the soil or becomes involved in groundwater and runoff. Eventually, this water reaches the oceans where rivers empty into them.

B. *The Carbon Cycle:* Carbon is found in all organic compounds. The basic carbon cycle involves the assimilation of carbon (as CO_2) by plants during photosynthesis and the return of carbon to the atmosphere by respiration and decomposition (again, as CO_2). There is a seasonal fluctuation in the carbon cycle due to winter in the Northern Hemisphere (much less carbon is fixed in the winter time). Carbon also flows into aquatic systems by combining with water to form carbonic acid (H_2CO_3) and in turn, bicarbonate ion. The amount of carbon present in aquatic ecosystems (essentially the oceans) is about 50 times as great as the amount present in the atmosphere. Some carbon is removed from the carbon cycle when dead organisms are turned into coal, oil, or peat. Humans are affecting the global carbon cycle by burning large amounts of otherwise unavailable carbon when we use fossil fuels. The CO_2 released in this way is increasing the CO_2 concentration in the atmosphere.

C. *The Nitrogen Cycle:* Nitrogen is a fundamental component of proteins. Most nitrogen is directly assimilated from the environment by plants. Nitrogen (as N_2) makes up about 80% of the earth's atmosphere but this gas is an unavailable source of nitrogen for plants. To obtain nitrogen in a usable form plants need nitrate (NO_3^-). They obtain this in two ways. A family of plants called Legumes has established a mutualistic relationship with nitrogen-fixing bacteria that lets them obtain nitrogen directly from atmospheric nitrogen. Most other plants obtain (NO_3^-) via the nitrification of ammonium ion, which itself comes from decomposition by fungi and bacteria (through ammonification) and the direct fixation of nitrogen gas by some bacteria. Consumers obtain nitrogen from producers and other consumers depending on the trophic level. Nitrogen gas' is returned to the atmosphere by denitrifying bacteria.

Although nitrogen cycles between the atmosphere and living things, most nitrogen is recycled locally through assimilation and decomposition.

III. Human Influence on Ecosystems

A. *Poisons in Food Chains:* Humans create a lot of different chemical compounds that are released intentionally or by accident into the environment. Some of these are toxic. In other cases, action by microorganisms in the environment create toxic compounds from them. Some of these compounds can persist for years in the environment and end up being stored in the tissues of the organisms that eat them.

For example, two classes of compounds containing carbon-chlorine bonds are particularly pernicious. Organochlorine insecticides (*e.g.*, DDT) and PCBs (polychlorinated biphenyls) were produced in vast quantities after World War II. Both these compounds are fat-soluble and so they have accumulated in the fatty tissue of the organisms that have ingested them. Animals that eat the organisms that contain these compounds can themselves be contaminated, often with very high doses. The increasing concentrations of organochlorines and PCBs that are found in animals that are at ever higher levels in the food chain is known as biological magnification. For example, DDT might be present in ca. 0.000003 ppm (parts per million) in a given water sample but might be as high as 25 ppm by the time it reaches the top carnivore in a food chain. Humans are not immune to this phenomenon. There is probably no one on earth today who is not carrying at least some of these chemicals in their fatty tissue. These chemicals can even be passed on to newborns via the milk of nursing mothers.

B. Carbon Dioxide Emissions and the Greenhouse Effect- an alteration of the carbon cycle. Modern industrial society is based largely on the burning of fossil fuels for energy. Fossil fuels are sources of hydrocarbons that were formed from the decayed remains of organisms millions of years ago. They include oil and coal. When these are burned for their energy CO_2 is released as a by-product. The level of CO_2 in the atmosphere has gradually increased over time due, apparently, to this release of CO_2 by human activity. This increasing CO_2 concentration has been correlated with an increase in average global temperature. CO_2 is responsible for retaining heat in our atmosphere. The more of it there is, the warmer the earth will be. If the average temperature of the earth increases enough (and his need only be ca. 3-4 degrees C), large-scale melting of the polar ice caps could result with the consequent flooding of many low lying areas, including many of world's most populous cities. The earth is undeniably warming, but scientists are still uncertain about the degree to which this is human influenced.

C. *Depletion of the Ozone Layer:* One product of free oxygen in the atmosphere is ozone (O_3). Ozone is harmful in high concentrations when we breathe it, but high in the atmosphere it serves a protective role by absorbing UV light, which is very energetic and harmful to living things. The amount of ozone in the atmosphere is decreasing, probably as the result of reactions with human made chemicals called chlorofluorocarbons. These are chemicals that had until recently been used as refrigerants and propellants in aerosol cans. These chemicals can react with ozone to destroy it. "Ozone holes" have been documented at high latitudes and these grow and shrink seasonally. There are now production bans on these chemicals in many parts of the world. A thinning of the ozone would mean more ultraviolet light exposure for living things and a potential increase in mutation levels.

D. *Human-caused Extinctions:* Extinctions are a normal part of life on planet earth but seldom has the rate of extinction been as great as it is now, and it is primarily due to human activity. As human population increases so does its demand on the world's resources. One of these resources is space and humans are encroaching on the natural world to a greater degree each year. That encroachment means habitat loss for many species and as habitat is lost, many species go extinct. In some places, like the rain forests, species are being lost so quickly that the rate might be as high as 1000-10,000 times what it normally is.

Biodiversity Hotspots: Biodiversity is not evenly distributed on the planet. Some areas harbor far greater concentrations of living creatures than others. Scientists had identified those biologically rich areas under the greatest threat of destruction as biodiversity hotspots, and are concentrating conservation in these areas to minimise it. Approximately 25 priority hotspots have been identified based on three criteria: the number of species present, the number of those species found exclusively in an ecosystem and the degree of threat they face.

History of the Hotspots: British ecologist Norman Myers created the biodiversity hotspots concept in two scientific papers published in 1.988 and 1990. Myers recognised that a modest number of hotspot ecosystems covering a small total land area, most often in tropical forest areas, accounted for a high percentage of global biodiversity.

Areas of the Hotspots

1. Tropical Andes
2. Mediterranean Basin
3. Madagascar/Indian Ocean Islands
4. Mesoamerica
5. Caribbean Islands
6. Indo-Burma
7. Atlantic Forest of Brazil
8. Philippines
9. Cape Floristic Region of South Africa
10. Mountains of South Central China
11. Sundaland
12. Brazilian Cerrado
13. Southwest Australia
14. Polynesia and Micronesia
15. New Caledonia
16. Choco/Darien/Western Ecuador
17. Western Ghats and Sri Lanka
18. California Floristic Province

19. Succulent Karoo
20. New Zealand
21. Central Chile
22. Guinean Forests of West Africa
23. Caucasus
24. Eastern Arc Mountains, Coastal Forests of Kenya and Tanzania
25. Wallacea

Some facts about biodiversity:

- Diversity increases from polar zones to tropics which is true for almost all taxa.
- Most species found on the earth are in tropical rainforests.
- Tropical rain forests cover 6% of land area and contain 50% of all species.
- Diversity is high in the tropics due to favourable Productivity, Habitat heterogeneity, Niche specialisation, Population interactions, Climate history.
- About 60 million years ago - most of Earth was tropical forest. Since Cenozoic era there has been a continuous cooling and drying of global climate which led to evolution of new biomes - temperate forests, grasslands, deserts, etc.
- Lineages in tropical forests have been evolving for more than 60 million years- while lineages that are adapted to cooler and drier conditions are younger.
- Ice ages in northern latitudes during last 2 million years depleted diversity available at that time.
- Each biome within each biogeographic realm has endemic species - species found nowhere else.
- Now humans have modified 60% of land surface.
- Habitat fragmentation leads to extinction.
- 99.9% of all species found at some time on earth have gone extinct.

- The background extinction rate of species is 1/million species/year.
- Given a current species diversity of 1 - 10 million, we should see 1-10 extinctions per year. But the current extinction rate is much higher - *e.g.*, birds. Of 10,000 species of birds (1-10% of total number of species) approximately 2,000 became extinct in last 2,000 years. There is good evidence that birds are "typical" in their risk of extinction—there are as more extinctions in other taxa.
- It is estimated that the current rate of extinction is around 17,000 - 100,000 extinctions per year. An average estimate is 35,000/year which is similar to past mass extinctions.
- Leading cause of extinction is tropical deforestation.
- Rate of tropical deforestation is 1-3%. At this rate 50% of all species present on earth may be lost by 2050.

Successful action to conserve biodiversity must address the full range of causes of its current loss and embrace the opportunities that genes, species, and ecosystems provide for sustainable development. Because the goal of biodiversity conservation-supporting sustainable development by protection and using biological resources in ways that do not diminish the world's variety of genes and species or destroy important habitats and ecosystems is so broad, any biodiversity conservation strategy must also have a broad scope. A biodiversity campaign can be broken down into three basic elements:

- saving biodiversity;
- studying it; and
- using it sustainably and equitably.

A. *Saving Biodiversity:* Saving biodiversity means taking steps to protect genes, species, habitats, and ecosystems. The best way to maintain species is to maintain their habitats. Saving biodiversity therefore often involves efforts to prevent the degradation of key natural ecosystems and to manage and protect them effectively. But since many of the world's habitats have been modified for such human uses as agriculture, the programme must include measures to maintain diversity on lands and in waters that have already been disturbed. A third component is restoring lost species to their former habitats and preserving species in gene-banks, zoos, botanic gardens, and other off-site (ex situ) facilities.

B. *Studying Biodiversity:* Studying biodiversity means documenting its composition, distribution, structure, and function; understanding the roles and functions of genes, species, and ecosystems; grasping the complex links between modified and natural systems; and using this understanding to support sustainable development. It also means building awareness of biodiversity's values, providing opportunities for people to appreciate nature's variety, integrating biodiversity issues into educational curricula, and ensuring that the public has access to information on biodiversity, especially on developments that will influence it locally.

C. *Using Biodiversity Sustainably and Equitably:* Using biodiversity sustainably and equitably means husbanding biological resources so that they last indefinitely, making sure that biodiversity is used to improve the human condition, and seeing that these resources are shared equitably. "Use" does not, however, automaticeally imply consumption. Often, the best economic use of biodiversity may be to maintain it in its natural state for its ecological or cultural values, as in the cases of forested watersheds or sacred groves.

Biodiversity Conservation Scheme: The biodiversity conservation agenda must encompass much more than concern for protected areas, threatened species, zoos or seed-banks and its constituency must be broad-based.

It has to take place within the wide context of the move toward sustainable living discussed in *Our Common Future* the report of the World Commission on Environment and Development-and detailed in *Caring for the Earth,* the successor and complement to the *World Conservation Strategy.*

How can biodiversity conservation be addressed within the context of sustainable development, as it must succeed? There must be new contacts and partnerships within communities, bringing biologists and resource managers together with social scientists, political leaders, businessmen, religious leaders, farmers, journalists, artists, planners, teachers, and lawyers.

There must be dialogue between central and local governments, industry, and citizen's groups, including non-governmental environment and development organisations, and women's and indigenous people's organisations. New mechanisms for discussion, negotiation, and common action are all essential.

Biodiversity conservation must take place at the individual level, the global level, and in between. Effective conservation efforts begin in the fields, forests, watersheds, grasslands, coastal zones, and settlements where people live and work. But complementary governmental efforts are needed to address the many facets of biodiversity conservation beyond the capacity of local communities, or involving resources that are of national importance. By the same token, international cooperation is essential, given the global nature of the biodiversity crisis and the lack of national resources in many countries.

Many essential elements of biodiversity conservation require sustained commitment, but will not show immediate results. Policies, institutions, laws, and attitudes do not change overnight. Expanding human capacity, carrying out first-rate research, and conducting biodiversity inventories take time and money ana may have no immediate pay-off. But they create the larger context in which enduring change can take hold and emergency measures have at least a hope of succeeding. Still, immediate action is needed. Irreplaceable genes, species, and ecosystems are disappearing at a rate unprecedented in human history, and essential development is at risk as a result. Immediate action is needed

- to defend these threatened living resources;
- to reform the policies that invite such losses;
- to conduct inventory and study of resource use in key ecosystems and countries;
- to monitor changes and impeding threats;
- to better manage threaten protected areas;
- to mobilise funding; and
- to support national and grassroot's conservation initiatives.

Making a Biodiversity Strategy: The limited conservation resources available must be focussed strategically on opportunities likely to yield the greatest conservation benefits, and five key strategic objectives offer significant possibilities for effective action:

1. To develop national and international policy frameworks that foster the sustainable use of biological resources and the maintenance of biodiversity;
2. To create conditions and incentives for effective conservation by local communities;
3. To strengthen and apply moie broadly the tools for conserving biodiversity;
4. To greatly strengthen the human capacity for conserving and using biodiversity sustainably, particularly in developing countries;
5. To catalyze conservation action through international cooperation and national planning.

Promoting Actions for Biodiversity Conservation

- Ratify the Convention on Biological Diversity like other international treaties.
- Establish a mechanism such as an International Panel on Biodiversity Conservation (preferably within the Convention on Biological Diversity), including scientists, non-governmental organisations, and policy-makers to provide guidance on priorities for the protection, understanding, and sustainable and equitable use of biodiversity.
- Establish an Early Warning Network, linked to the Convention on Biological Diversity, to monitor potential threats to biodiversity and mobilise action against these threats.
- Integrate biodiversity conservation in national planning processes.

Hydro-electric Power: Hydro-electric power plants convert the kinetic energy contained in falling water into electricity. The energy in flowing water is ultimately derived from the sun, and is therefore constantly being renewed. Energy contained in sunlight evaporates water from the oceans and deposits it on land in the form of rain. Differences in land elevation result in rainfall runoff, and allow some of the original solar energy to be captured as hydro-electric power. Hydro power is currently the world's largest renewable source of electricity, accounting for 6% of worldwide energy supply or about 15% of the world's electricity. Traditionally thought of as a cheap and clean source of electricity, most large hydro-electric schemes being planned today are coming up against a great deal of opposition from environmental groups and local people.

The first recorded use of water power was a clock, built around 250 BC. Since that time, humans have used falling water to provide power for grain and saw mills, as well as a host of other applications. The first use of moving water to produce electricity was a waterwheel on the Fox river in Wisconsin in 1882, two years after Thomas Edison unveiled the incandescent light bulb. Early hydro-electric power plants were much more reliable and efficient than the fossil fuel fired plants of the day. This resulted in a proliferation of small to medium-sized hydro-electric generating stations distributed wherever there was an adequate supply of moving water and a need for electricity.

As electricity demand soared in the middle years of this century, and the efficiency of coal and oil-fueled power plants increased, small hydro plants fell out of favour. Most new hydro-electric development is focused on huge "mega-projects".

The majority of these power plants involved large dams which flooded vast areas of land to provide water storage and therefore a constant supply of electricity. In recent years, the environmental impacts of such large hydro projects are being identified as a cause for concern. It is becoming increasingly difficult for developers to build new dams because of opposition from environmentalists and people living on the land to be flooded. This is shown by the opposition to projects such as Tehri and Narmada projects.

Hydro-electric Power Plants: Hydro-electric power plants capture the energy released by water falling through a vertical distance, and transform this energy into useful electricity. In general, falling water is channeled through a turbine which converts the water's energy into mechanical power. The rotation of the water turbines is transferred to a generator which produces electricity. The amount of electricity which can be generated at a hydro-electric plant is dependent upon two factors.

(1) the vertical distance through which the water falls, called the "head", and

(2) the flow rate, measured as volume per unit time.

The electricity produced is proportional to the product of the head and the rate of flow. The following is an equation which may be used to roughly determine the amount of electricity which can be generated by a potential hydro-electric power site:

$$POWER (kW) = 5.9 \times FLOW \times HEAD$$

In this equation, FLOW is measured in cubic meters per second and HEAD is measured in meters. Based on the facts presented above, hydro-electric power plants can generally be divided into two categories. "High head" power plants are the most common and generally utilise a dam to store water at an increased elevation.

The use of a dam to impound water also provides the capability of storing water during rainy periods and releasing it during dry periods. This results in the consistent and reliable production of electricity, able to meet demand. Heads for this type of power plant may be greater than 1000 feet. Most large hydro-electric facilities are of the high head variety High head plants with storage are very valuable to electric utilities because they can be quickly adjusted to meet the electrical demand on a distribution system.

"Low head" hydro-electric plants are power plants which generally utilise heads of only a few meters or less. Power plants of this type may utilise a low dam or weir to channel water, or no dam and simply use the "run of the river". Run of the river-generating stations cannot store water, thus their electric output varies with seasonal flows of water in a river. A large volume of water must pass through a low head hydro plant's turbines in order to produce a useful amount of power. Hydro-electric facilities with a capacity of less than about 25 MW (1 MW = 1,000,000 Watts) are generally referred to as "small hydro", although hydro-electric technology is basically the same regardless of generating capacity.

"Pumped Storage" is another form of hydro-electric power. Pumped storage facilities use excess electrical system capacity, generally available at night, to pump water from one reservoir to another reservoir at a higher elevation. During periods of peak electrical demand, water from the higher reservoir is released through turbines to the lower reservoir, and electricity is produced. Although pumped storage sites are not net producers of electricity—it actually takes more electricity to pump the water up than is recovered when it is released—they are a valuable addition to electricity supply systems. Their value is in their ability to store electricity for use at a later time when peak demands are occurring. Storage is even more valuable if intermittent sources of electricity such as solar or wind are hooked into a system.

Impact Environmental by Hydro-electric Power Plants: Hydro-electric power plants have many environmental impacts, some of which are just beginning to be understood. These impacts, however, must be weighed against the environmental impacts of alternative sources of electricity. Until recently there was an almost universal belief that hydro power was a clean and environmentally safe method of producing electricity. Hydro-electric power plants do not emit any of the standard atmospheric pollutants such as carbon dioxide or sulphur dioxide given off by fossil fuel fired power plants. In this respect, hydro power is better than burning coal, oil or natural gas to produce electricity, as it does not contribute to global warming or acid rain. Similarly, hydro-electric power plants do not result in the risks of radioactive contamination associated with nuclear power plants.

A few recent studies of large reservoirs created behind hydro dams have suggested that decaying vegetation, submerged by flooding, may give off quantities of greenhouse gases equivalent to those from other sources of electricity. If this turns out to be true, hydro-electric facilities that flood large areas of land might be significant contributors to global warming. Run of the river hydro- plants without dams and reservoirs would not be a source of these greenhouse gases.

The most obvious impact of hydro-electric dams is the flooding of vast areas of land, much of it previously forested or used for agriculture. The size of reservoirs created can extremely be large. Reservoirs can be used for ensuring adequate water supplies, providing irrigation, and recreation; but in several cases they have flooded the homelands of native peoples, whose way of life has then been destroyed. Many rare ecosystems are also threatened by hydro-electric development.

Large dams and reservoirs can have other impacts on a watershed. Damming a river can alter the amount and quality of water in the river downstream of the dam, as well as preventing fish from migrating upstream to spawn.

These impacts can be reduced by requiring minimum flows downstream of a dam, and by creating fish ladders which allow fish to move upstream past the dam. Silt, normally carried downstream to the lower reaches of a river, is trapped by a dam and deposited on the bed of the reservoir. This silt can slowly fill up a reservoir, decreasing the amount of water which can be stored and used for electrical generation.

The river downstream of the dam is also deprived of silt, which fertilizes the river's flood plain during high water periods. Bacteria present in decaying vegetation can also change mercury, present in rocks underlying a reservoir, into a form, which is soluble in water.

The mercury accumulates in the bodies of fish and poses a health hazard to those who depend on these fish for food. The water quality of many reservoirs also poses a health hazard due to new forms of bacteria, which grow in many of the hydro rivers. Therefore, run of the river type hydro plants generally have a smaller impact on the environment.

Significant Functions of Energy

The theoretical size of the worldwide hydropower is about four times greater than that which has been exploited at this time. The actual amount of electricity,' which will ever be generated by hydropower, will be much less than the theoretical potential. This is due to the environmental concerns outlined above and economic constraints. Much of the remaining hydro potential in the world exists in the developing countries of Africa and Asia.

Harnessing this resource would require billions of dollars, because hydroelectric facilities generally have very high construction costs. In the past, the World Bank has spent billions of foreign aid dollars on huge hydroelectric projects in the third world. Opposition to hydro power from environmentalists and native people, as well as new environmental assessments at the World Bank will restrict the amount of money spent on hydroelectric power construction in the developing countries of the world.

In North America and Europe, a large percentage of hydropower potential has already been developed. Public opposition to large hydro schemes will probably result in very little new development of big dams and reservoirs. Small scale and low-head hydro capacity will probably increase in the future as research on low-head turbines, and standardised turbine production, lowers the costs of hydroelectric power at sites with low-heads.

New computerised control systems and improved turbines may allow more electricity to be generated from existing facilities in the future. As well, many small hydro-electric sites were abandoned in the 1950's and 60's when the price of oil and coal was very low, and their environmental impacts unrealised. Increased fuel prices in the future could result in these facilities being refurbished.

Hydro-electric power has always been an important part of the world's electricity supply, providing reliable, cost-effective electricity, and will continue to do so in the future. Hydro power has environmental impacts which are very different from those of fossil fuel power plants.

The actual effects of dams and reservoirs on various ecosystems are only now becoming understood. The future of hydro-electric power will depend upon future demand for electricity, as well as how societies value the environmental impacts of hydro-electric power compared to the impacts of other sources of electricity.

Comparing Energy's Alternative Sources

Solar

Advantages: Always available, no pollution

Disadvantages: Low efficiency (15%); Very high initial costs; lack of adequate storage materials (batteries); High cost to the consumer.

Hydro

Advantages: No pollution; Very high efficiency (80%); little waste heat; low cost per KWH; can adjust KWH output to peak loads; recreation money can be generated.

Disadvantages: Fishes are endangered, Sediment buildup and dam failure; changes watershed characteristics; alters hydrological cycle.

Wind

Advantages: Supplemental power in windy areas; best alternative for individual home owner Disadvantages: Highly variable source; relatively low efficiency (30%).

Geothermal

Advantages: Very high efficiency; low initial costs.

Disadvantages: Non-renewable (more is taken out than can be put in by nature); highly local resource.

Ocean Thermal Energy Conversion

Advantages: Enormous energy; steady flow for decades; can be used on large scale; exploits natural temperature gradients in the ocean.

Disadvantages: Enormous engineering effort; extremely high cost; Damage to coastal environments possible.

Tidal Energy

Advantages: Steady source; energy extracted from the potential and kinetic energy of the earth sun-moon system; can exploit bore tides for maximum efficiency.

Disadvantages: Low duty cycle due to intermittent tidal flow; huge modification of coastal environment; very high costs for low duty cycle source.

Nuclear Energy: A debated energy source, of future.

Alternative Sources of Energy with Respect to their Environmental Impact: To combat global warming and the other problems associated with fossil fuels, we must switch to renewable energy sources like sunlight, wind, and biomass. All renewable energy technologies are not appropriate to all applications or locations and must be chosen after due consideration of all the factors involved. However, as with conventional energy production, there are environmental issues to be considered.

Energy from Wind: It can be taken as a source of energy most benign to the environment. It produces no air or water pollution, involves no toxic or hazardous substances (other than those commonly involved with large machines), and poses no threat to public safety.

One of the most misunderstood aspects of wind power is its use of land. Most studies assume that wind turbines will be spaced a certain distance apart and that all of the land in between should be regarded as occupied. This leads to some quite disturbing estimates of the land area required to produce substantial quantities of wind power. According to one widely circulated report from the 1970s, generating 20 per cent of US electricity from windy areas in 1975 would have required siting turbines on 18,000 square miles, or an area about 7 per cent the size of Texas.

In reality, however, the wind turbines themselves occupy only a small fraction of this land area, and the rest can be used for other purposes or left in its natural state. For this reason, wind power development is ideally suited to farming areas. In Europe, farmers plant right up to the base of turbine towers, while in California cows can be seen peacefully grazing in their shadow. The leasing of land for wind turbines, far from interfering with farm operations, can bring substantial benefits to landowners in the form of increased income and land values. Perhaps the greatest potential for wind power development is in the coastal areas like Tamil Nadu, Gujarat, etc. In other settings, however, wind power development can create serious land-use conflicts.

In forested areas it may mean clearing trees and cutting roads, a prospect that is sure to generate controversy, except possibly in areas where heavy logging has already occurred. And near populated areas, wind projects often run into stiff opposition from people. In California, bird deaths from electrocution or collisions with spinning rotors have emerged as a problem in a wind "farm,". In appropriate areas, and with imagination, careful planning, and early contacts between the wind industry, environmental groups, and affected communities, siting and environmental problems can be easily solved.

Solar Energy: Since solar power systems generate no air pollution during operation, the primary environmental, health, and safety issues involve how they are manufactured, installed, and ultimately disposed of Energy is required to manufacture and install solar components, and any fossil fuels used for this purpose will generate emissions.

Thus, an important question is how much fossil energy input is required for solar systems compared to the fossil energy consumed by comparable conventional energy systems. Although this varies depending upon the technology and climate, the energy balance is generally favourable to solar systems in applications where they are cost-effective, and it is improving with each successive generation of technology. According to some studies, for example, solar water heaters increase the amount of hot water generated per unit of fossil energy invested by at least a factor of two compared to natural gas water heating and by at least a factor of eight compared to electric water heating.

Materials used in some solar systems can create health and safety hazards for workers and anyone else coming into contact with them. In particular, the manufacturing of photovoltaic cells often requires hazardous materials such as arsenic and cadmium. Even relatively inert silicon, a major material used in solar cells can be hazardous to workers if it is breathed in as dust. Workers involved in manufacturing photovoltaic modules and components must consequently be protected from exposure to these materials.

None of these potential hazards is much different in quality or magnitude from the innumerable hazards people face routinely in an industrial society. Through effective regulation, the dangers can very likely be kept at a very low level.

The large amount of land required for utility-scale solar power plants-approximately one square kilometer for every 20-60 megawatts (MW) generated-poses an additional problem, especially where wildlife protection is a concern. But this problem is not unique to solar power plants. Generating electricity from coal actually requires as much or more land per unit of energy delivered if the land used in strip mining is taken into account. Solar-thermal plants (like most conventional power plants) also require cooling water, which may be costly or scarce in desert areas.

Large central power plants are not the only option for generating energy from sunlight, however, and are probably among the least promising. Because sunlight is dispersed, small-scale, dispersed applications are a better match to the resource. They can't take advantage of unused space on the roofs of homes and buildings and in urban and industrial lots.

Geothermal Energy: Geothermal energy is heat contained below the earth's surface. The only type of geothermal energy that has been widely developed is hydrothermal energy, which consists of trapped hot water or steam. However, new technologies are being developed to exploit hot dry rock (accessed by drilling deep into rock), geopressured resources (pressurised brine mixed with methane), and magma.

The various geothermal resource types differ in many respects, but they raise a common set of environmental issues. Air and water pollution are two leading concerns, along with the safe disposal of hazardous waste, siting, and land subsidence.

Since these resources would be exploited in a highly centralised fashion, reducing their environmental impacts to an acceptable level should be relatively easy. But it will always be difficult to site plants in scenic or otherwise environmentally sensitive areas due to strong public opposition.

The method used to convert geothermal steam or hot water to electricity directly affects the amount of waste generated. Closed-loop systems are almost totally benign, since gases or fluids removed from the well are not exposed to the atmosphere and are usually injected back into the ground after giving up their heat. Although this technology is more expensive than conventional open-loop systems, in some cases it may reduce scrubber and solid waste disposal costs enough to provide a significant economic advantage.

Open-loop systems, on the other hand, can generate large amounts of solid wastes as well as noxious fumes. Metals, minerals, and gases leach out into the geothermal steam or hot water as it passes through the rocks. The large amounts of chemicals released when geothermal fields are tapped for commercial production can be hazardous or objectionable to people living and working nearby.

In large geothermal development sites, steam vented at the surface contains hydrogen sulphide (H_2S)-accounting for the area's "rotten egg" smell-as well as ammonia, methane, and carbon dioxide. At hydrothermal plants carbon dioxide is expected to make up about 10 per cent of the gases trapped in geopressured brines. For each kilowatt-hour of electricity generated, however, the amount of carbon dioxide emitted is still only about 5 per cent of the amount emitted by a coal- or oil-fired power plant.

Scrubbers reduce air emissions but produce a watery sludge high in sulphur and vanadium, a heavy metal that can be toxic in high concentrations. Additional sludge is generated when hydrothermal steam is condensed, causing the dissolved solids to precipitate out. This sludge is generally high in silica compounds, chlorides, arsenic, mercury, nickel, and other toxic heavy metals. One costly method of waste disposal involves drying it as thoroughly as possible and shipping it to licensed hazardous waste sites. Research under way at New York points to the possibility of treating these wastes with microbes designed to recover commercially valuable metals while rendering the waste nontoxic.

Usually the best disposal method is to inject liquid wastes or redissolve the solids to pump them back into a porous stratum of a geothermal well. This technique is especially important at geopressured power plants because of the sheer volume of wastes they produce each day. Wastes must be injected well below fresh water aquifers to make certain that there is no communication between the usable water and waste-water strata. In addition to providing safe waste disposal, injection may also help prevent land subsidence. At Wairakei, New Zealand, where wastes and condensates were not injected for many years. This resulted in sinking of an area by 7.5 meters since 1958.

Most geothermal power plants will require a large amount of water for cooling or other purposes. In places where water is in short supply, this need could raise conflicts with other users for water resources.

The development of hydrothermal energy faces a special problem. Many hydrothermal reservoirs are located in or near wilderness areas of great natural beauty such as Yellowstone National Park in United States and in Laddakh in India. Proposed developments in such areas in USA have aroused intense opposition.

Biomass: Biomass power, derived from the burning of plant matter, raises more serious environmental issues than any other renewable resource except hydropower. Combustion of biomass and biomass-derived fuels produces air pollution; beyond this, there are concerns about the impacts of using land to grow energy crops.

How serious these impacts are will depend on how carefully the resource is managed. The picture is further complicated because there is no single biomass technology, but rather a wide variety of production and conversion methods, each with different environmental impacts.

Cause of Air Pollution because of use of Biomass as Fuel: Inevitably, the combustion of biomass produces air pollutants, including carbon monoxide, nitrogen Oxides, and particulates such as soot and ash. The amount of pollution emitted per unit of energy generated varies widely by technology, with wood-burning stoves and fireplaces generally the worst of fenders. Modern, enclosed fireplaces and wood stoves pollute much less than traditional, open fireplaces for the simple reason that they are more efficient. Specialised pollution control devices such as electrostatic precipitators (to remove particulates) are available, but their use will increase the production costs.

Emissions from conventional biomass-fueled power plants are generally similar to emissions from coal-fired power plants, with the notable difference that biomass facilities produce very little sulphur dioxide or toxic metals (cadmium, mercury, and others). The most serious problem is their particulate emissions, which must be controlled with special devices. More advanced technologies, such as the gasifier/combustion turbine combination, should generate much lower emissions, perhaps comparable to those of power plants fueled by natural gas.

Facilities that burn raw municipal waste present a unique pollution-control problem. This waste often contains toxic metals, chlorinated compounds, and plastics, which generate harmful emissions.

Since this problem is much less severe in facilities burning Refuse-Derived Fuel (RDF) -pelletised or shredded paper and other waste with most inorganic material removed most waste-to-energy plants built in the future are likely to use this fuel. Co-firing RDF in coal-fired power plants may provide an inexpensive way to reduce coal emissions without having to build new power plants.

Using biomass-derived methanol and ethanol as vehicle fuels, instead of conventional gasoline, could substantially reduce some types of pollution from automobiles. Both methanol and ethanol evaporate more slowly than gasoline, thus helping to reduce evaporative emissions of Volatile Organic Compounds (VOCs), which react with heat and sunlight to generate ground-level ozone (a component of smog).

According to US EPA estimates, in cars specifically designed to burn pure methanol or ethanol, VOC emissions from the tail pipe could be reduced 85 to 95 per cent, while carbon monoxide emissions could be reduced 30 to 90 per cent. However, emissions of nitrogen oxides, a source of acid precipitation, would not change significantly compared to gasoline-powered vehicles.

Some studies have indicated that the use of fuel alcohol increases emissions of formaldehyde and other aldehydes, compounds identified as potential carcinogens. Others counter that these results consider only tailpipe emissions, whereas VOCs, another significant pathway of aldehyde formation, are much lower in alcohol-burning vehicles. On balance, methanol vehicles would therefore decrease ozone levels. Overall, however, alcohol-fueled cars will not solve air pollution problems in dense urban areas, where electric cars or fuel cells represent better solutions.

Greenhouse Gases and Biomass Energy: A major benefit of substituting biomass for fossil fuels is that, if done in a sustainable fashion, it would greatly reduce emissions of greenhouses gases. The amount of carbon dioxide released when biomass is burned is very nearly the same as the amount required to replenish the plants grown to produce the biomass. Thus, in a sustainable fuel cycle, there would be no net emissions of carbon dioxide, although some fossil-fuel inputs may be required for planting, harvesting, transporting, and processing biomass. Yet, if efficient cultivation and conversion processes are used, the resulting emissions should be small (around 20 per cent of the emissions created by fossil fuels alone). And if the energy needed to produce and process biomass came from renewable sources in the first place, the net contribution to global warming would be zero.

Similarly, if biomass wastes such as crop residues or municipal solid wastes are used for energy, there should be lesser or no net greenhouse gas emissions. This will also reduce the release of methane from landfills.

Implications for Agriculture and Forestry: One effect of growing trees and other plants for energy is that it could benefit soil quality and farm economies. Energy crops could provide a steady supplemental income for farmers in off-seasons or allow them to work unused land without requiring additional equipment. Moreover, energy crops could be used to stabilise cropland or rangeland prone to erosion and flooding. Trees would be grown for several years before being harvested, and their roots and leaf litter could help stabilise the soil. The planting of coppicing, or self-regenerating, varieties like eucalyptus, poplar, etc. would minimise the need for disruptive tilling and planting. Perennial grasses harvested like hay could play a similar role.

If improperly managed, however, energy farming could have harmful environmental impacts. Although energy crops could be grown with less pesticide and fertilizer than conventional food crops, large-scale energy farming could nevertheless lead to increases in chemical use simply because more land would be under cultivation. It could also affect biodiversity through the destruction of species habitats, especially if forests are more intensively managed. If agricultural or forestry wastes and residues were used for fuel, then soils could be depleted of organic content and nutrients unless care was taken to leave enough wastes behind. These concerns point out the need for regulation and monitoring of energy crop development and waste use.

Energy farms may present a perfect opportunity to promote low-impact sustainable agriculture, or, as it is sometimes called, organic farming. Such methods could be adapted to energy farming. Nitrogen-fixing crops could be used to provide natural fertilizer, while crop diversity and use of pest parasites and predators could reduce pesticide use. Though such practices may not produce as high a yield as more intensive methods, this penalty could be offset by reduced energy and chemical costs. Increasing the amount of forest wood harvested for energy could have both positive and negative effects.

On one hand, it could provide an incentive for the forest-products industry to manage its resources more efficiently, and thus improve forest health. But it could also provide an excuse, under the "green" mantle, to exploit forests in an unsustainable fashion. Unfortunately, commercial forests have not always been soundly managed, and many people view with alarm the prospect of increased wood cutting. Their concerns can be met by tighter government controls on forestry practices and by following the principles of "excellent" forestry. If such principles are applied, it should be possible to extract energy from forests indefinitely.

Hydropower: The development of hydropower has become increasingly problematic due to strict considerations being given to adverse environmental impacts. The biggest obstacle in construction of large dams is that most suitable undeveloped sites are very often reserve forest, or inhabited by people. The Silent Valley project in Kerala was forced to be abandoned by public opposition. In United States for example, a series of large facilities on the Columbia River in Washington was forced to reduce its peak output by 1,000 MW to save an endangered species of salmon. Salmon numbers have declined rapidly because the young are forced to make a long and arduous trip downstream through several power plants, carrying risk of death from turbine blades at each stage. To ease this trip, hydropower plants may be required to divert water around their turbines at those times of the year when the fish attempt the trip.

That environmental concerns would constrain hydropower development more and more difficult in future though these plants produce no air pollution or greenhouse gases. Yet, as the salmon example makes clear, they affect the environment. The impact of very large dams is so great that many developed countries had dropped the plans to construct very large dams though they are being constructed in many developing countries like India and China due to their lucrative economics. The reservoirs created by such projects frequently inundate large areas of forest, farmland, wildlife habitats, scenic areas, and even towns.

In addition, the dams can cause radical changes in river ecosystems both upstream and downstream. Small hydropower plants using reservoirs can cause similar types of damage, though obviously on a smaller scale.

Some of the impacts on fish can be mitigated by installing "ladders" or other devices to allow fish to migrate over dams, and by maintaining minimum river-flow rates; screens can also be installed to keep fish away from turbine blades. In one case, flashing underwater lights placed in the Susquehanna River in Pennsylvania direct night-migrating American shad around turbines at a hydroelectric station. As environmental regulations have become more stringent, developing cost-effective mitigation measures such as these is essential. It is also a fact that hydropower is almost certainly approaching the limit of its potential in most of the developed countries.

Earth Formation

Most soils are composed of small particles of weathered rock that are silicate minerals. These minerals consist of polymeric structures in which the fundamental unit is a silicon atom surrounded tetrahedrally by four oxygen atoms, the latter in turn are each bonded to another silicon, and so on, the resulting structure is an extended network. There are variations in the structure and H^+ Na^+, K^+, Mg^{2+}, Ca^{2+}, Al^{3+} Fe^{2+} may be in the structure.

The weathering of the silicate minerals from rocks can involve chemical reactions such as the substitution by aluminium for some of the silicon atoms. Eventually such reactions yield clay minerals which are a class of soil materials. Clay minerals have a particle size less than 2 mm. Small colloidal clay particles possess an outer layer of cations which are bound electrostatically to an electrically-charged inner layer.

Depending on the concentration of cations in the water surrounding the clay particle, the cations on the particle are capable of being exchanged, *e.g.* if the soil is acidic, the metal ions on the surface will be displaced by H^+ ions and the metal ions will then enter the aqueous phase.

Besides minerals soil contains organic matter, water and air in varying concentrations. The organic matter, which gives soil its dark colour is primarily a material called humus, which is undecomposed plant material. It contains mostly protein and lignin, both of which are mostly insoluble in water. Lignin contains polymeric strands of benzene rings connected by chains of carbon and oxygen atoms. Due to partial oxidation of some of the lignin, many of the resulting polymeric strands contain Carboxylic acid groups, - COOH; this portion of humus is known as humic and fulvic acids, and is soluble in alkaline solutions.

The ultimate sink for heavy metals, and for many toxic organic compounds as well, is deposition and burial in soil and sediments. Heavy metals often accumulate in the top layer of the soil, and are therefore accessible for uptake by the roots of crops.

Humic materials have a great affinity for heavy metal cations, and extract them from the water that passes through them. This happens mostly because of the formation of complexes with the metal ions by - COOH groups in the humic and fulvic acids.

Humic acid complexes are normally water - insoluble and smaller fulvic acid complexes are water-soluble.

Heavy metals are retained by soil in three ways:

(a) By adsorption onto the surfaces of mineral particles
(b) By complexation by humic substances in organic particles
(c) By precipitation reactions.

Like many other chemicals, heavy metals often are adsorbed onto the surfaces of particulates - especially organic ones—suspended in water. In lakes such particles eventually settle to the bottom of the lakes and are buried when other sediments accumulate on top of them. Sometimes such sink mechanism doesn't work due to desorption of chemicals or consumption of particles by bottom-growing organisms or due to anaerobic bacterial activities. Ex. Anaerobic bacteria methylate the mercuric ion to form $(CH_3)2Hg$ and CH_3HgX, which then rapidly desorb from sediment particles and dissolve in water, and enter in food chain. The biogeochemical cycle of bacterial methylation and demethylation of mercury in sediments.

Analysing the Risk: The objective of risk analysis is to provide a scientific framework to help decision makers and other concerned individuals in making judicious decisions that will ultimately solve or mitigate health and environmental problems. Risk analyses are used in many situations like

(a) Analyzing potential health and environmental problems resulting from exposure to chemicals
(b) Selecting sites for hazardous-chemical-producing units

(c) Comparing alternate technologies available

(d) Analyzing potential risks due to fire and explosion hazards.

To be meaningful and cheap risk analyses are usually based on a relatively narrow set of hazards *i.e.*, they focus on a particular problem, for example, cancer and chronic health risks while not considering others like ecological problems.

The terms and their definitions used in the field of risk analysis are given below:

Hazard: Hazard refers to the source of risk. For example a toxic chemical is a hazardous to human health but is not a risk unless someone is exposed to it.

Hazard Identification: It is the first step in risk analysis. It is the process of reconising all the hazards with the potential to harm an individual or the environment.

Risk: It is the probability of exposure of individuals, populations or ecosystems to toxic substances or to hazardous conditions. Risks can be lowered by reducing any of its components.

(a) The probability of risk taking place

(b) The probability of exposure

(c) The severity of consequences.

Risk Assessment: It is the second step in risk analysis. It is an estimate of the severity and likelihood of harm to populations or ecosystems from exposure to a hazard. It includes

(a) Assessment of sources

(b) Assessment of exposure

(c) Assessment of dose-response

(d) Risk characterisation, which calculates a numerical estimate of the risk.

Risk management: It is the effective use of a risk assessment. It may include evaluation of alternatives, informed decisions and the follow through activities.

Analysing the Risk: Risk analysis includes four steps:

(a) hazard identification

(b) risk assessment

(c) risk management

(d) risk communication

Risk analyses use science, engineering, probability and statistics to estimate and evaluate the magnitude and likelihood of health and environmental risks.

(a) *Hazard Identification:* It establishes the existence of hazards at specific site. Hazards may be toxic substances or specific conditions that have the potential to be harmful to human health or the environment.

(b) *Source:* For risk analysis further information like, toxicity, rate of discharge, conditions of release (temperature and pressure), etc. are required to decide a worst case scenario.

(c) *Exposure Assessment:* It determines the identification of the toxic species, frequency of exposure, duration of exposure and the routes of exposure to human populations and the environment. Exposure assessment data is obtained from actual monitoring, mathematical models, or reasonable scientific estimates.

(d) *Dose-response Assessment:* This is also called toxicity assessment. It involves

(i) identification of the adverse health effects due to chemical exposure

(ii) relationship between the magnitude of the exposure and adverse effects.

The response to an exposure may be

(i) no observable effect

(ii) temporary effects, like eye irritation

(iii) permanent injury like liver damage

(iv) chronic functional impairment like emphysema of lungs

(v) death

(e) *Risk Characterisation:* In it the results of exposure assessment and dose-response are used to determine if the current or future levels of exposure are potentially problematic. Toxicology and exposure results are combined to give qualitative and quantitative expressions of risk. Risk characterisation includes the probability of health and environmental effects.

(f) *Risk Management:* It uses all the informations discussed above, together with other technical, business, environmental, social and political information to

(i) evaluate alternative

(ii) determine the best action to reduce or eliminate risk.

(iii) design a plan to implement this action.

(g) *Risk Communication:* The results of a risk analysis must be communicated to technical and public communities to gain benefits. The purpose of an effective risk analysis is to assist decision makers, companies and citizens to work on things that matter most and ignore things that matter the least.

Desertification: It is a process where by the productive potential of arid and semi and land falls by 10% or more. Moderate desertification is 10-25% drop in productivity, severe desertification is a 2550% drop, and very severe desertification is a drop of 50% or more, generally creating huge gullies and sand dunes. Since last 50 years, an estimated 8.1 million square kilometers have become desertified. This threatens the livelihoods of at least 900 million people in 100 countries. Every year low-to-moderate desertification affects 60,000 square kilometer area, and another 210,000 square kilometer area undergo severe desertification and loose so much soil and fertility that they are no longer useful for farming or grazing.

Reasons for Desertification

1. Overgrazing on fragile arid and semi and rangelands.
2. Deforestation without reforestation.
3. Surface mining without land reclamation.
4. Build-up of salt and water logging of soil.
5. Farming on land with unsuitable terrain or soil.
6. Soil compaction by heavy farm machinery and cattle hoof.

- *Desertification in hilly regions of the Himalayas:* Deforestation in Himalayas, involving clearance of natural forests and plantation of monocultures like *Pinus roxburghi, Eucalyptus camaldulensis,* etc. have upset the ecosystem by changing various soil (edaphic) and biological properties. Nutrient cycling has become poor, original rich germplasm is lost and the area is invaded by exotic weeds. These areas are not able to recover and are losing their fertility. The entire west Khasi hill district of Meghalaya in North-east Himalayas, Laddakh and parts of Kumaon and Garhwal are now facing the serious problem of desertification.
- *Disappearing Tea gardens in Chhota Nagpur:* This hilly region used to be a good forested area towards the turn of the century and used to receive fairly frequent afternoon showers favouring tea plantations. Following the destruction of forests, rainfall declined in Chhota Nagpur to such an extent that tea-gardens also disappeared from the region.
- *Waning Rainfall in Udhagamandalam (Ooty):* The sub-normal rainfall during 1965-84 at Ooty in Nilgiri mountains has been found to be closely associated with declining forest cover in this region in the past 20 years. The rainfall pattern was found to fluctuate with wooded land area in the hills. When the Nilgiri mountains had luxuriant forest cover annual rainfall used to be much higher.

Sardar Sarovar Dam (Uprooted Forests And Tribals)

A Case Study: The dam is situated on river Narmada and is spread over three states of Gujarat, Maharashtra and Madhya Pradesh. Although the project is aimed at providing irrigation water, drinking water and electricity to the three states, the environmental impacts of the project have raised challenging questions.

A total of 1,44,731 ha of land will be submerged by the dam, out of which 56,547 ha is forest land. A total of 573 villages are to be submerged by the Narmada Dam.

Submergence of about 40,000 ha of forest under Narmada Sagar, 13,800 ha under Sardar Sarovar and 2,500 ha under Omkareshwar would further create pressure on remaining forest areas in adjoining areas. Submergence area is very rich in wildlife *e.g.* tigers, panthers, bears, wolves, pangolins, hyenas, jackals, flying squirrels, antelopes, black bucks, chinkara, marsh crocodiles, turtles, etc. Many of these species are listed in schedule I & II of Wildlife Protection Act, 1972. Thus massive loss of these wildlife species is apprehended, due to the devastation of the forest under the project.

As per the estimates of the Institute of Urban Affairs, New Delhi, the Narmada valley project will lead to eventual displacement of more than one million people, which is probably the largest rehabilitation issue ever encountered as per the World Bank. Uprooting of the tribals and their forced shifting in far-flung areas may not be easily adjusted to.

Besides serious economic deprivation, the displacement will affect the tribal peoples' culture, their beliefs, myths and rituals, festivals, songs and dances, all closely associated with the hills, forest and streams.

Most of these tribals belong to poor, under privileged schedule castes and tribes who are being uprooted from a place where they have lived for generations. The displaced persons have to undergo hardship and distress for the sake of development and prosperity of a larger section of the society. It is therefore the duty of the project proponents and government to pay maximum attention for proper rehabilitation of the displaced tribals.

Case Studies

- ***Mining and Quarrying in Udaipur***

About 200 open cast mining and quarrying centers in Udaipur, about half of which are illegal are involved in stone-mining including soapstone, building stone, rock phosphate and dolomite. The mines spread over 15,000 hectares in Udaipur have caused many adverse impacts on environment. About 150 tonnes of explosives are used per month in blasting.

The overburden, wash-off, discharge of mine water, etc. pollute the water. The Maton mines have badly polluted the Ahar river. The hills around the mines are devoid of any vegetation except a few scattered patches and the hills are suffering from acute soil erosion. The waste water flows towards a big tank of "Bag Dara". Due to scarcity of water people are compelled to use this effluent for irrigation purpose.

The blasting activity has adversely affected the fauna and the animals like tiger, lion, deer and even hare, fox, wild cats and birds have disappeared from the mining area.

- ***Mining in Sariska Tiger Reserve in Aravallis***

The Aravalli range is spread over about 692 km in the North-west India covering Gujarat, Rajasthan, Haryana and Delhi. The hill region is very rich in biodiversity as well as mineral resources. The Sariska tiger reserve has gentle sloppy hills, vertical rocky valleys, flat plains as well as deep gorges. The reserve is very rich in wildlife and has enormous mineral reserves like quartzite, Schists, marble and granite in abundance. Mining operations within and around the Sariska Tiger reserve has left many areas permanently infertile and barren. The precious wildlife is under serious threat. We must preserve the

Aravalli series as a National Heritage and the Supreme Court on December 31st, 1991 has given a judgement in response to a Public Interest Litigation of Tarun Bharat Sangh, an NGO wherein both Centre and State Government of Rajasthan have been directed to ensure that all mining activity within the park be stopped. More than 400 mines were shut immediately. But, still some illegal mining is in progress.

- ***Uranium Mining in Nalgonda, A.P.—The public hearing***

The present reserves of Uranium in Jaduguda mines, Jharkhand can supply the yellow cake only till 2004. There is a pressing need for mining more uranium to meet the demands of India's nuclear programme. The Uranium Corporation of India (UCIL) proposes to mine uranium from the deposits in Lambapur and Peddagattu villages of Nalgonda district in Andhra Pradesh and a processing unit at about 18 kms at Mallapur. The plan is to extract the ore of 11.02 million tons in 20 years.

The IUCL is trying its best to allure the villagers through employment opportunities. But, experts charge the company for keeping silence on the possible contamination of water bodies in the area. The proposed mines are just 1 km from human habitation and hardly 10 km from Nagarjun Sagar Dam and barely 4 km from the Akkampalli reservoir which is Hyderabad's new source of drinking water.

It is estimated that 20 years of mining would generate about 7.5 million metric tones of radioactive waste of which 99.9% will be left behind. The villagers are very likely to be affected by the radioactive wastes. Though IUCL claims that there won't be any such accidents, but no one can deny that it is a highly hazardous industry and safety measures cannot be overlooked. The pathetic condition of Jaduguda Uranium mines in Jharkhand where there is a black history of massive deaths and devastation have outraged the public, who don't want it to be repeated for Nalgonda. The proposed mines would cover about 445 ha of Yellapurum Reserve Forest and the Rajiv Gandhi Tiger Sanctuary. The public hearing held just recently in February, 2004 witnessed strong protests from NGOs and many villagers. The fate of the proposed mining is yet to be decided.

Case Studies: Salinity and water logging in Punjab, Haryana and Rajasthan:

> The first alarming report of salt-affected wasteland formation in connection with irrigation practices came from Haryana (then Punjab) in 1858. It was reported that several villages in Panipat, Rohtak and Delhi lying in command area of Western Yamuna Canal were suffering from destructive saline efflorescence. The "Reh Committee" in 1886 drew the attention of the government on some vital points showing a close relationship between irrigation, drainage and spread of "reh" and "usar" soils.
>
> The floods of 1947, 1950, 1952, 1954-56 in Punjab resulted in aggravated water logging with serious drainage problems. Introduction of canal irrigation in 1.2 m ha in Haryana resulted in rise in water-table followed by water-logging and salinity in many irrigated areas causing huge economic losses as a result of fall in crop productivity.
>
> Rajasthan too has suffered badly in this regard following the biggest irrigation project "Indira Gandhi Canal Project" and the sufferings of a big area in Western Rajasthan have changed from a condition of "water-starved wasteland" to that of a "water soaked wasteland".

Case Study: A build-up of DDT concentration: A striking case of biomagnification of DDT (a broad range insecticide) was observed when some birds like Osprey were found to suffer a sharp decline in their population.

The young ones of these birds were found to hatch out in

premature condition leading to their death. This was later found to be due to bio-magnification of DDT through the food chain. DDT sprayed for pest control was in very low concentration, but its concentration increased along the food chain through phytoplanktons to zooplanktons and then to fish which was eaten by the birds.

The concentration of DDT was magnified several thousand times in the birds which caused thinning of shells in the birds' eggs, causing death of the young ones. It becomes very clear from the above instance that the animals occupying the higher trophic levels are at a greater risk of biomagnification of toxic chemicals. Human beings consuming milk, eggs and meat are at a higher trophic level. So, we have to stop indiscriminate use of pesticides and heavy metals if we wish to save ourselves from their biologically-magnified toxic levels.

7

Gradual Growth

Systematic Cultivation

In the beginning of the nineteenth century people with varying degrees of scientific background (medical men, naturalists and engineers) started coming to India from Europe, at first in search of services under local princes and chieftains and later on to man the scientific surveys and establishments set up by the government. These men framed in European institutions and laboratories, spent the best part of their lives in India and left an excellent record of their work in various branches of science.

Despite such long contacts, introduction of modern science in India was extremely slow. There were many causes of this delayed reaction. The European scientific world in India was limited to field sciences and not to basic sciences that depended on mathematics and laboratory work.

Indian fauna and flora attracted the attention of European naturalists from the seventeenth century. Modern zoological researchers in India had their beginnings in random and scattered observations by the naturalists on elephants, fishes, serpents, molluscs, birds and mammals.

The Colonial Government adopted from the beginning a policy of secrecy and exclusion of Indians from Government scientific work. For technological jobs not considered of military importance, restrictions against the employment of Indians were relaxed as they could be employed very cheaply. As an example of the low salaries to trained and qualified Indians/ the order of the Governor-General dated 28th January 1835, establishing the Calcutta Medical College, regulated that "as inducement for pupils of a respectable class to enter the Institution, the pay of the Native Doctors, who shall have been educated at the College, and have received the certificates of qualifications, shall be Rs. 30.00 p.m. rising to Rs. 50.00 p.m. after 14 years of service, whereas an European Assistant to the Superintendent of the college shall draw a staff salary of Rs. 600.00 p.m. in addition to his registered pay and allowances." Calcutta Medical College established in 1835 under a number of capable teachers, became an important institution, for the study of anatomy, physiology and medicine (along with surgery), as well as of chemistry, botany and natural philosophy.

Indians asked for Western Education in Science from the Government. The Colonial power agreed to create facilities but with miserly financial provision just sufficient to train a few clerks able to operate in English in Government offices and European Commercial houses, and not to enable the native people to pick up treasures from European science.

So the hard way of working for science began. Mahendra Lal Sircar *(born* in 1833), an M.D. of the Calcutta Medical College, clearly saw that science would never develop and strike deep roots in this country under foreign tutelage and that Indians themselves must come forward to raise funds and found institutions for the training of scientists and organisation of research.

In 1876 Dr. Sircar himself founded India's first research institution, "The Indian Association for the Cultivation of Science," completely under Indian Management and control with finances derived from private subscriptions. In the first half of the twentieth century the association developed into an important centre for research in the physical sciences—in optics, acoustics, scattering of light, X-rays and magnetism; and C.V. Raman performed his Nobel Prize Winning Experiments in its unassuming laboratories, *i.e.* western India. Jamshedji Tata a businessman, prepared plans for a similar institution for technical and scientific education and research which finally took the shape of "The Institute of Science" at Bangalore in the beginning of the twentieth century.

The Educational Despatch of 1854 paved the way for University education. In 1857 the three Universities of Calcutta, Bombay and Madras came into existence, but they came out just as examining bodies with powers to grant degrees. The actual teaching and academic work were left to the colleges. Fortified by the University Act of 1904, which empowered Universities to appoint professors and lecturers, to hold and manage educational endowments, and to erect, equip and maintain libraries, laboratories and museums, Asutosh Mookerji, Vice Chancellor of Calcutta University took the initiative in establishing the, first University College of Science. Undaunted by the Government's refusal to provide funds for the creation of professorships and other facilities, Asutosh obtained princely endowments from Sir Tarak Nath Palit and Sir Rash Behary Ghosh, who had amassed enormous amounts of money and property in the legal profession, and established a number of chairs in Chemistry, Physics, Applied Mathematics and Botany.

According to the terms of the endowment, professorships could be filled only by Indians, a clause which further irritated the Government firmly entrenched in the view that only Europeans were suited for such high positions. This University College of Science, although starved financially all through, produced a group of physicists and chemists who received international recognition for their scientific developments and institutions staffed by high salaried Europeans and maintained and patronised by Government funds cut a sorry figure.

The development of science in India was greatly accelerated after independence (August 1947). In 1950 the Government of India appointed a Planning Commission for preparing a blueprint of all-round economic development. In 1954 the Indian Parliament accepted socialism as a political goal. Declaring these objectives, fullest emphasis was laid on the development of science and technology on all fronts.

In 1957, the Government took one step further in adopting a National Science Policy Resolution that envisaged the cultivation of science and scientific research in all its aspects, assured an adequate supply, within the country, of research scientists of highest quality through an intensive programme of training, promised the availability of conditions and an atmosphere of academic freedom in which the creative talent of men and women would find full scope in scientific activity. The resolution thus reaffirmed the Government decision to encourage science and develop a healthy scientific community as a sound basis after a balanced economic development.

General Aspects

In our country Science is witnessed to be taught from very early days. If we look at the development of science in India, we find that very early in her civilisation India developed a great interest in mathematics and Ayurveda. During British Empire introduction of modern science in India was extremely slow, and the development of science in India was greatly accelerated after independence.

Before the rise of modern science in Europe, around the 17th century, the level of advancement in sciences (astronomy, mathematics, medicine, biology, metaphysics, etc.) achieved by ancient and medieval societies in the old world did not differ appreciably from one cultural area to another. In India the development of sciences is as old as her civilisation itself. Her peculiar geographical position enabled her to become the natural meeting ground of many nations and cultures and, in consequence, played an important role in the transmission and diffusion of ideas very early in her civilisation, India developed medical and alchemical practices of the Ayurveda. In mathematics, India developed a great interest and aptitude and made notable contributions to the number theory, the decimal place value, algorism, trigonometry and algebra.

New Advancements

India is now engaged in a broad spectrum of scientific research, both fundamental and applied, in Government, Universities and Private Research Establishments. In recent years there has been extraordinary success in developing new polymers, ceramics, composites, superconductors, nanomaterials, smart materials and biomaterials. Biotechnology, Genetic Engineering and Biomedical Research are some other fields in which India has started entering.

AIDS (Acquired Immune Deficiency Syndrome) poses a threat to India as a large number of people are infected with HIV (Human Immunodeficiency Virus). There are no drugs today for AIDS. The problem which has dogged anti HIV drugs is that resistant mutant forms of virus are formed within mere weeks. Lot of research work is to be done in search of a vaccine against HIV. Per capita consumption of energy in India is very low compared to developed countries, and even that we are unable to afford. In the years to come when we have already entered twenty-first century, we will need much more energy. The major effort in next decade would have to be through an increase in the production of coal and a search of new reserves of oil. We should also give due emphasis to new technologies for solar energy and hydrogen energy. India is also to expand its nuclear power programmes. Nuclear energy could play an important role in meeting our bulk energy requirement. Our expertise in the field of nuclear power technology as well as related research areas, is an asset for us. For all this we need huge amounts of funds, as scientific research has become costly. If we compare the funds available for scientific research in India compared to some other developing countries, India has a very gloomy picture. South Korea had planned to increase its R&D spending to over 5 per cent of its GDP by year 2000.

China had planned to increase its R&D investment from 0.5 per cent to 1.5 per cent of GDP by the year 2000. Unfortunately India's R&D expenditure has come down to 0.89 per cent of GDP from 1.1 per cent earlier. We need to realise that in order to be able to do competitive scientific research and development we have to bring our R&D expenditure to at least about 1.5 per cent of GDP and focus our efforts on a few selected programmes and projects.

Our scientists want people to support them and understand their needs, and they are to inform people about science, why science is needed, what they are doing and why. They (our scientists) are to build up partnership with public, industries, politicians and bureaucrats. They need informed friends of science at all places.

When changes which affect our future, are happening and will happen at such a rapid rate, and are based on science and technology, it is necessary that our scientists be more close to decision making.

In the existing state of scientific advancement and development of resources for research, the rate of scientific growth of developed countries is likely to continue to be faster than that of the developing countries like India. Science has become deeply involved with defence and big industry, with the result that big sciences like atomic energy, space, etc., has been and probably will remain concentrated in super powers, In nuclear, space, computer and few other sciences, developing nations like India are already at the mercy of super powers.

To be self-sufficient we are to change our science curriculum right from the school stage. We are not just to teach science but also scientific method. We are not just to teach science content but also science processes, the ways in which scientists advance their knowledge and solve problems.

Science should be presented to students as a way in which they can conduct an inquiry into the nature of things as well as a body of information built up by other people. The science processes are being neglected and the school science has been concerned almost exclusively with the content—the body of information. Science is not just content. Science is content plus science processes.

If we want to advance in science like other developed countries, science processes should be given due importance like science content in school science curriculum, and students should be encouraged to become personally involved in solving problems and in discovering some science for themselves.

Science and its Application

- In the beginning of twentieth century science was not a school subject in our country. The Report of the Secondary School Commission 1953, recommended the teaching of General Science as a compulsory subject in the high and higher secondary schools.
- The All India Seminar on the Teaching of Science in Secondary Schools (1956) dealt with almost all the problems facing the inclusion of General Science as a Core Subject for the Higher Secondary Classes—syllabus, apparatus, teaching aids, textbooks, science clubs, science museums, examination techniques etc. It suggested a uniform system of science teaching for the entire country, suited to its needs and resources.
- Indian Parliamentary and Scientific Committee (1962) studied the allied problems of science education in schools like:–growth of school population,–shortage of

qualified science teachers,–accelerated achievement in science,–demand for increase in technically trained manpower,–growing importance of science in the affairs of mankind, and–changes in the processes and goals of science.

- UNESCO Planning Commission (1963-64) worked on the problems of science education in India and suggested ways to improve it. As a follow up programme, Department of Science and Mathematics Education of the NCERT took up the pilot project of preparing new disciplinary science curricula at middle (VI-VIII) level -text books (Physics, Chemistry, Biology), teachers' guides, science kits, kit guides, teacher training films and evaluation material. This Disciplined Science Programme was changed to Integrated Science Programme in mid-eighties.
- In 1970 under UNICEF Assisted Science Education Programme (SEP) Primary "Science is Doing" Programme was developed by NCERT, which was used in primary schools throughout the country. This Programme was a package of classes I and II Science Syllabus, Class III-V Science Texts, Teachers Guides, Primary Science Kit, Kit Guide, and Teacher Training Films. This Programme was changed to EVS Programme in mid-eighties.
- Kothari Commission (1964-66) recommendations were implemented in 1975 when Science for All (SFA) was introduced as a part of general education during the first ten years of schooling. With this 10+2+3 education scheme started with an additional year of schooling, in the country. First Disciplined Science Course (Physics, Chemistry, Biology) was introduced at secondary level (IX-X). This was changed to Disciplined Science A-Course (Physics, Chemistry, Biology) and Integrated Science B-Course. Students had an option either to take Science A-Course or to take Science B-Course.

This again changed to just one Integrated Science Course for all. At +2, Senior Secondary level (XI-XII) Disciplined Science Course (3 different science subjects-Physics, Chemistry, Biology) started fróm the very beginning of 10+2+3 education scheme.

- Then in 1986 the National Education Policy Document (NPE-1986) came out. If some one asks, "What is new in this new education policy (NPE-1986), "Implementation" perhaps will be the right answer. Much emphasis has been given to quality pre-service and in-service teacher education in the policy document. For this District Institutes of Education (DIETs) and Institutes of Advance Studies in Education (LASEs) were established throughout the country for Elementary and Secondary Education respectively.

8

SAFETY OF PLANTS

- Plant Pathology or Phytopathology (phyton-plant; pathos-ailments (disease); logos-knowledge) is that branch of agricultural science which deals with the cause etiology, resulting losses and management of plant disease. The science of plant pathology has four major objectives:
  1. To study the living, non-living and environmental causes of disease or disorder in plants.
  2. To study the mechanism (s) of disease development by pathogens.
  3. To study the interaction between the plant and the pathogen in relation to the overall environment.
  4. Develop systems of management of the diseases and reduce the loss of caused by plant diseases.
- Its aim is not only to destroy the enemies of plants but also to increase productivity of the crop and profits of the farmers.

Diseases in Plants

Bacterial Diseases

	Disease	*Causal Organism*
1.	Ratoon stunning disease of sugarcane (R.S.D.S)	*Cleibactor xyli*
2.	Stalk rot of maize	*Erwinia chrysanthemi pv. zeae*
3.	Tuber soft rot	*E. cartovora*
4.	Common scab of potato	*Streptomyces scabis*
5.	Citrus greening disease	*Diaphorina citri*
6.	Leaf spot of mango	*Xanthomonas campestris* pv. *Mangifera indica*
7.	Bacterial spike blight	*Clavibacter tritici*
8.	Bacterial brown rot of potato	*Pseudomonas solanacearum*
9.	Bacterial leaf streak of rice	*Xanthomonas campestris pv. oryzicola*
10.	Bacterial leaf blight	*Xanthomonas campestris pv. oryzae*
11.	Citrus canker	*Xanthomonas campestris pv. citri*
12.	Angular leaf spot	*Xanthomonas campestris pv. malvacearum*

Viral Diseases

1.	Leaf curl of papaya	*Nicotiana virus-10*
2.	Papaya Mosaic	Potex virus group
3.	Fiji disease of sugarcane	Fiji virus
4.	Mosaic of sugarcane	Sugarcane Mosaic Virus (S.M.V.)
5.	Bunchy top of banana	Banana virus-1

Contd...

	Disease	*Causal Organism*
6.	Leaf roll of potato	Potato leaf roll virus (Luteo group of virus) *(IAS(Pre) 2002)*
7.	Mild mosaic of potato	Potato mild mosaic virus
8.	Yellow vein mosaic of okra	Yellow vein mosaic virus
9.	Rugose mosaic of potato	Potato virus x and potato virus *y*
10.	Tomato spotted wilt	Tomato spotted wilt virus (Tospo group of virus) *(IAS (Pre) 2002)*
11.	Rice Tungro	Rice tungro virus (Bacilliform virus) (Badna group) *(IAS (Pre) 2002)*
Nematode Diseases		
1.	Ear cockle of wheat	*Anguina tritici* *(HAS(Pre) 2003)*
2.	Molya disease of barley	*Heterodera avenae* *(PG. Bikaner 2000)*
3.	Root knot of vegetables crops	*Meloidogyne javanica* *M. incognita* *M. arenaria*
Fungal Diseases		
1.	Club root of crucifers	*Plasmodiophora brassicae*
2.	Black wart of potato	*Synchytrium endobioticum*
3.	Damping off of seedlings	Most common generas *Phythium* *Phytophthora* *Fusarium*

Contd...

	Disease	Causal Organism
		Rhizoctonia
		Botrytis
		Ozonium
4.	Brown spot disease of maize	*Physoderma zeae-maydis*
5.	Fruit rot of cucurbits	*Phythium aphanidermatum*
6.	Foot or stem rot of papaya	*P. aphanidermatum*
7.	White rust of crucifers	*Albugo candida*
8.	Green ear disease of bajra	*Sclerospora graminicola*
9.	Downy mildew of Jowar	*Sclerospora graminicola pr. andropopognis*
10.	Downy mildew of maize	*Pernosplerospora sacchari*
11.	Downy mildew of pea	*Pernospora pisi*
12.	Downy mildew of cucurbits	*Psedopernospora cubensis*
13.	Downy mildew of crucifers	*Pernospora parasitica* (PG 2000)
14.	Downy mildew *of* grape vines	*Plasmospora viticola*
15.	Soft rot of ginger	*P. myriotylum*
16.	Root rot of turmeric	*Pythium aphanidermatum*
17.	Late blight of potato	*Phytophthora infestance*
18.	Blight of Colocasia	*Phytophthora colocasiae*
19.	Seedling blight of castor	*Phytophthqra parasitica*
20.	Bud rot of palm	*Phytophthora palmivora*
21.	Gumosis of citrus	*Phytophthora palmivora and P. parasitica*
22.	Blight of sesamum	*P. parasitica var. sesami*
23.	Stem galls of coriander	*Promyces macrosporus*

Contd...

	Disease	*Causal Organism*
24.	Peach leaf curl	*Taphrina deformis*
25.	Leaf spot of turmeric	*Taphrina maculans*
26.	Soft rot of apples	*Penicillium expansum*
27.	Powdery mildew of pea	*Erysiphi polygoni*
28.	Powdery mildew of cereals	*Erysiphe graminis*
29.	Powdery mildew of cucurbits	*E. cichoiacearum*
30.	Powdery mildew of grape vine	*Uncincola necator*
31.	Powdery mildew of apple	*Podasphara leucotricha*
32.	Ring spot of sugarcane	*Leptospharia sacchari*
33.	Stem rot of paddy	*Sclerotium oryzae*
34.	Ergot of cereals	*Calviceps purpuria*
35.	Ergot of bajra	*Calviceps fusiformis*
36.	False smut of rice	*Claviceps oryzae sativae*
37.	Apple scap	*Venturia inacqualis*
38.	Stem canker of fruit	*Necteria cinnabarinatrees*
39.	Stem rot of papaya	*Sclerotinia sclerotiorum*
40.	Loose smut of wheat	*Ustilago tritici U. nuda tritici (PG 2000)*
41.	Brown rust of wheat	*Puccinia recondita*
42.	Yellow rust of wheat	*P. striiformis*
43.	Black or stem rust of wheat	*P. graminis tritici*
44.	Flag smut of wheat	*U. agropyri*
45.	Leaf smut of rice	*Entyloma oryzae*
46.	Bunt of rice	*Neovossia horrida*

Contd...

	Disease	Causal Organism
47.	Karnal bunt of wheat	*Tilletia tritici*
48.	Stinking smut of wheat	*Tilletia caries*
49.	Smut of sorghum	*Tolyposporium ehrenbergii*
50.	Smut of bajra	*T. penicillariae*
51.	Head smut of sorghum	*Sphacelotheca reiliana*
52.	Grain smut of sorghum	*Sphacelotheca sorghi*
53.	Smut of maize	*Ustilago maydis*
54.	Covered smut of barley	*Ustilago hordi*
55.	Loose smut of bajra	*Ustilago nuda* (JRF1999)
56.	Smut of sugarcane	*Ustilago scitaminea*
57.	Covered smut of oats	*Ustilago kalleri*
58.	Loose smut of oats	*U. avenae*
59.	Bean rust	*Uromyces phoseoli typica*
60.	Cowpea rust	*U. phaseoli vignae*
61.	Leaf rust of coffee	*Hemelia vastatrix* and *Hemellia coffcicola*
62.	Rust of linseed	*Melampsora lini*
63.	Early blight of potato	*Alternaria solani*
64.	Alternaria leaf spot of crucifers	*Alternoria brassicae*
65.	Leaf blight of wheat	*Alternaria triticina*
66.	Leaf spot of groundnut Tikka disease	*Cercasporidium personatum* or *C. arachidicola* (PC 2000)
67.	Leaf spot of rice	*Hèlminthosporium oryzae*

Contd...

	Disease	***Causal Organism***
68.	Blast of rice	*Pyricularia oryzae*
69.	Red rot of sugarcane	*Colletotrichum falcatum*
70.	Die back of chillies	*Colletotrichum capsici*
71.	Mango anthracnose	*Colletotrichum gloeosporioides*
72.	Anthracnose of bean	*Colletotrichum lindemuthianum*
73.	Wilt of pigeon pea	*Fusarium udum (PC 2000)*
74.	Wilt of cotton	*Fusarium oxysporium sp. vasinfectum*
75.	Wilt of linseed	*F. oxysporium (f. lini)*
76.	Panama wilt of banana	*F. oxysporium (f. cubense)*
77.	Wilt disease of sugarcane	*Cephelosporium sacchari*
78.	Stalk rot	*Fusarium monilform*
79.	Blight of gram	*Ascochyta rabiei*
80.	Charcoal rot of soyabean	*Macrophomina phaseolina*
81.	Sheath blight of rice	*Rhizoctonia solani*
82.	Black scurf of potato	*Rhizoctonia solani*

Mycoplasmal Diseases

1.	Little leaf of brinjal	Mycoplasma
2.	Purple top roll of potato	MLOs
3.	Marginal flavescence	MLOs
4.	Witches broom	MLOs *(JRF2002)*
5.	Grassy shoot of sugarcane	MLOs
6.	Sesamum phyllody	MLOs

Contd...

	Disease	Causal Organism
	Physiological, Non-infectious or Non-parasitic Disease	
1.	Black heart of potato	Unfavourable O_2 condition during storage
2.	Black tip or mango necrosis	Polluted gases (SO_2)
3.	Tip burn of paddy	Reduced O_2 and N_2 under submerged condition
4.	Khaira disease of rice	Zn deficiency *(JRF1999;JRF2002)*
5.	Grey speck of oat	Mn deficiency
6.	Heart rot of beets	Lack of Mn
7.	Pahala blight of sugarcane	Mn deficiency *(IAS 2000)*
8.	White bud of maize	Lack of Zn
9.	Mottle leaf of citrus	Zn deficiency *(PCS 2002)*
10.	Whiptail *in* cauliflower	Mo deficiency *(JRF2001)*
11.	Browning or hollow stem of cauliflower	Boron deficiency
12.	Top sickness of tobacco	deficiency of B
13.	Yellow spot of citrus	deficiency of Mo

Important Glossary

- *Abiotic Disease:* A disease due to non-parasitic cause.
- *Alternate Host:* One of the two kinds of plants on which a parasitic fungus must develop to complete the life cycle. A host of a pathogen growing during off season for the main host crop.
- *Antibiotic:* Chemical compound (Metabolite) of an organism the inhibits of other micro-organism..

- *Atrophy:* Under development or wasting away of a body part.
- *Autoecious Fungus:* A parasitic fungus that completes its entire life cycle on the same host.
- *Axenic:* That type of culture of an organism in which no other organisms are present.
- *Antheridium:* Male sex organ in fungi.
- *Bactericide:* A chemical that kills bacteria.
- *Bactericcin:* Bactericidal substance produced by certain strain of bacteria and active against one or more strains of the same or closely related species, their production is genetically controlled by extrachromosomal DNA.
- *Bacteriophage:* A virus that infects bacteria and usually destroys them.
- *Biological Control:* Total or partial destruction of population of a plant pathogen by use of other organism.
- *Biotechnology:* The use of genetically modified organisms and or modern techniques and processes with biological systems for disease control, crop production.
- *Blight:* Symptoms characterised by rapid killing of leaves, flowers and stems.
- *Blister:* A raised lesion on the leaf surface which opens to expose spores.
- *Blotch:* A disease characterised by large irregular spots or blots on leaves, shoots and stems.
- *Canker:* A necrotic, often sunken, lesion on stem, branch, twig or fruit.
- *Chlorosis:* Yellowing of normally green tissue due to destruction of chlorophyll or failure of chlorophyll formation.
- *Chronic Disease:* A disease which persists in the host for a long period of time.
- *Chronic Symptoms:* Symptoms that appear over a long time.

- *Contagious Disease*: A parasitic disease whose pathogen can spread from plant to plant by contact.
- *Circulative Viruses*: Viruses that are acquired by their vectors through their mouth parts, accumulate internally, then are passed through their tissues and introduced into new plants via. the mouth part of the vector.
- *Damage*: Any reduction in the quantity and/or quality of yield that results from injury.
- *Damping Off*: Seed decay in soil or seedling or blight usually caused by soil inhabiting fungi.
- *Diagnosis*: Identification of the nature and cause of the disease.
- *Die-back*: A symptoms of disease characterised by progressive death of shoots branches and roots from the tip backward.
- *Disease Assessment*: Quantitative measurement of disease or the process of measuring disease quantitatively.
- *Disease Intensity*: General term for amount of disease present in a population.
- *Disease Prevalence*: Generally incidence of fields with diseased plants in a defined geographic area.
- *Downy Mildew*: A plant disease in which the sporangio or conidiophores and spores of the fungus appear as downy growth on the host surface.
- *ELISA (Enzyme-Linked Immuno Sorbent Assay)*: A serological test in which one antibody carries with it an enzyme that releases a coloured compound.
- *Epidemic:* A widespread and severe outbreak of a disease in a large population.
- *Epidemology:* The study of factors affecting outbreak and spread of an infectious disease.

- *Eradicant:* A chemical that destroys a pathogen at its source.
- *Eradication:* Control of plant diseases by eliminating the pathogen after it is established or by eliminating the plant that carries the pathogen.
- *Etiology:* Study of science of the cause of the plant disease and nature of the causal agent.
- *Exanthema:* Eruption or discharge of gum or other substances from diseased tissues.
- *Facultative Parasite:* An organism that is usually a saprophyte but which under some conditions may become a parasite.
- *Facultative Saprophyte:* An organism which spends most of its life as a parasite but may become a saprophyte for sometime under certain conditions.
- *Fumigant:* A volatile toxic substance or gas used to disinfest certain areas or plants.
- *Fumigation:* The application of a fumigant for disinfestation.
- *Fungicide:* A compound that kills fungi.
- *Gall:* A swelling or overgrowth produced on a plant as a result of infection by certain pathogens.
- *Gummosis:* Production of gum by or in a plant tissue.
- *Homothallism:* The phenomenon in which a fungus produces compatable male and female gametes on the same mycelium.
- *Horizontal Resistance:* Partial resistance equally effective against all races of a pathogen.
- *Heterothallic Fungi:* Fungi producing compatable male and female gametes on physiologically distinct mycelia.
- *Hyaline:* Colourless, transparent.
- *Hypovirulence:* Reduced virulence of a pathogen stain

as a result of the presence of transmissible double stranded RNA.

- *Infection:* The establishment of parasitic relationship of a parasite with the host.
- *Injury:* Damage to a plant by animal, physical or chemical agent visible or measurable symptoms and/or signs caused by a pathogen or pest.
- *Integrated Control:* An approach to plant disease management that uses all available methods of control of a disease or of all or most of the diseases or pests of a crop plant for best control results but with the least cost and the least damage to the environment.
- *In Vitro:* In culture, outside the host.
- *In Vivo:* In the host.
- *Intracellular:* Within the cells through the cells.
- *Intercellular:* Between the cells.
- *Leaf Spot:* A self limiting lesion on the leaf.
- *Local Lesion:* A localised spot produced on a leaf upon mechanical inoculation with a virus.
- *Latent Virus:* A virus that does not induce symptom development in the host.
- *Latent Infection:* The state in which a host is infected but does not show symptoms.
- *Mycoplasma Like Organism (MLO):* The organism that other wise looks and behave like a mycoplasma but has not yet been grown in artificial culture.
- *Mottle:* An irregular pattern of indistinct light and dark areas.
- *Mosaic:* Symptoms of certain virus diseases of plants characteried by intermingled patches of normal and light green or yellowish colour.
- *Molt:* To cast off the cuticle; Eedysis instar.

- *Mold:* Any profuse or woolly fungus growth on damp or decaying matter or no surfaces of plant disease.
- *Mildew:* A fungal disease of plants in which the mycelium and spores of the fungus are seen as a whitish growth on the host surface. When the growth appears like a downy growth it is called downy mildew.
- When it is like a white or dull powder it is known as powdery mildew.
- *Necrotic:* Dead and discoloured.
- *Nematicide:* A chemical compound or physical agent that kills nematodes.
- *Non-infectious Disease:* A disease that is caused by an environmental factor, not by organisms and viruses.
- *Obligate Parasite:* A parasite that in nature can grow and multiply only on a living organism.
- *Oomycetes:* A fungus that forms oospores.
- *Parasite:* An organism that lives on or in another living organism and draws its nutrition from that organism.
- *Pathogen:* An entity that can incite disease.
- *Polygenic:* A character controlled by many genes.
- *Proliferation:* Abnormal outgrowth.
- *Phytotoxic:* Tox to plants.
- *Pathogenicity:* The ability of a pathogen to cause disease.
- *Postule:* A small blister-like elevation of the epidermis, often opening to expose spores.
- *Quarantine:* Control of import and export of plants or plant parts to prevent spread of disease and pest.
- *Resistance:* Ability of a host plant to supress or retard the activity of a pathogen or either injurious factors.
- *Resistant:* Possessing qualities that hinder the development of a given pathogen, little or no infection.
- *Ring Spot:* A disease symptom characterised by a

circular area of chlorosis with a green centre often caused by viruses.

- *Rosette:* Short, bunchy habit of plant growth.
- *Rot:* The softening discolouration and often disintegration of a succulant plant tissue as a result of fungal or bacterial infection.
- *Rust:* A disease giving a rusty appearance to a plant and caused by one of the uredinales.
- *Scab:* A roughened crust-like disease area of the surface of a plant organ. A disease in which such areas are formed.
- *Scorch:* "Burning" of leaf margins as a result of infection or unfavourable environmental conditions.
- *Secondary Infection:* Infection caused by inoculum produced as a result of primary infection.
- *Serology:* A method using specificity of the antigen-antibody reaction for the detection and identification of antigenic substances and the organisms that carry them.
- *Smut:* A disease caused by smut fungi and characterised by appearance of shooty powdery mass on the affected organs.
- *Sooty Mold:* A shooty coating on foliage and fruits formed by the dark hyphae of some fungi that live on the honeydew secreted by insects.
- *Sporadic:* Term applied to a disease that breaks out occasionally without being constantly destructive.
- *Spore:* A single to many cell reproductive body of fungi and some lower plants which may develop into a new plant.
- *Stem-pitting:* A symptom of some virus diseases of trees characterised by depressions on the stem under the bark.
- *Sterile Fungi:* Fungi in which no form of spore has been seen.
- *Susceptibility:* The inability of a plant to resist the effect of pathogen or other injurious factors.

- *Symptom:* The external and internal reactions of the host to invasion of a pathogen alterations in the plant due to infection; signs of infection.
- *Syndrome:* A group of signs and symptoms that occur together and characterise the disease.
- *Systemic:* Spreading throughout the plant body internally.
- *Tolerance:* The ability of a plant to sustain the effects of a disease without dying or suffering serious injury or crop loss.
- *Toxicity:* The capability of a compound to produce injury.
- *Tumour:* And uncontrolled overgrowth of tissue or tissues.
- *Vertical Resistance:* Complete resistance to some races of the pathogen but not to other.
- *Virion:* A complete virus particle.
- *Viroids:* Small, naked, low molecular weight ribonucleic acid (RNA) that can infect plants, replicate, themselves and cause disease.
- *Virus:* A submicroscopic obligate parasite consisting of only nucleic acid and coat protein.
- *Virulent:* Capable of causing a severe disease strongly pathogenic highly aggressive.
- *Wilt:* Loss of rigidity and dropping of plant parts generally caused by insufficient water in plant body or by infection resulting in blockage of water transport or toxicity.
- *Witches Broom:* Broom like overgrowth or massed proliferation caused by dense clustering of branches of a plant.
- *Yellows:* A plant disease characterised by yellowing and stunting of the host plant.
- *Zoospores:* Spores bearing flagella and capable of movement in water.

Table

Crop	*Disease or Insect*	*Fungicide or Insecticide Comman Name*	*Fungicide or Insecticide Trade Name*	*Remarks*
Barley, Oats, Rye, Wheat	Seed decay and seeding blights including the scab fungus (seed treatments do dot control the head blight phase of scab disease)	Captan	Captan 30-DD Captan 300 Captan 400 Captan 400 DD	Thiram formulations are not registered for oats. Vitavax formulations, Raxil-Thiram, add Terra-coat LT-2 are dot registered for rye. Use according to instructions on label. Do dot use treated seed for feed or food.
		Thiram	Thiram-50WP Gustafson 42-S	
		Corboxin	Vitavax	Not for use on rye
		carboxin + ca tad	Vitavax 20-20	
		carboxin + thiram	Vitavax 200	
		difenconazole + mefanoxam	Dividend XL RTA 5-10 fl oz/100	For wheat use only. Apply pounds of seed.
		difenconazole + mefanoxam	Dividend Apron XL 0.5-1.0 fl oz/100	For wheat use only. Apply pounds of seed.
		PCNB	Gustafson LT-2N	Not for use on rye
		tebuconazole + thiram	Raxil-Thiram	

Contd...

Barley	Loose smut, Covered smut	carboxin	Vitavax Vitavax 34 Vitavax 75 W	Use according to instructions on label. Must be applied as a slurry treatment for complete coverage.
		carboxin + ca tad carboxin + thiram tebuconzole + thiram	Vitavax 20-20 Vitavax 200 Raxil-Thiram	
	Covered smut	PCNB	Terra-Coat LT-2	Use according to instructions on the label. Should be applied as a slurry treatment for complete coverage.
	Barley stripe	imazalil	Gustafson Flo-Pro IMZ	Use according to label instructions
Oats	Loose smut, covered smut	carboxin carboxin carboxin + thiram	Vitavax Vitavax 34 Vitavax 200	Use according to instructions on label. Must be applied as a slurry treatment for complete coverage.
	Loose smut	PCNB + tebuconazole + thiram	Terra-Coat Lt-2 Raxil-Thiram	Use according to instructions on the label. Should be applied as a slurry treatment for complete coverage.

Contd...

Wheat	Loose smut	Carboxin	Vitavax Vitavax 34	Use according to instructions on label. Must be applied as slurry treatment for complete coverage.
		Carboxin + ca tan	Vitavax 20-20	
		carboxin + thiram	Vitavax 200	
		difenconazole + mefanoxam	Dividend XL RTA	Apply 5-10 fl oz/100 Ib of seed.
		defenconazole + mefanoxam	Dividend Apron XL	Apply 0.5-1.0 fl oz/ 100 Ib of seed.
		tebuconazole + thiram	Raxil-Thiram	
		triadimen	Baytan 30 Flo%vable	This material is only registered for application by certified seed conditioners and is not available for farmer application. Rate is 0.75-1.5 fl oz/ 100 Ib of seed. Caution:Treated seed should not be planted late in the planting season or planted more 1.5 inches deep. Use according to label instructions.

Contd...

Wheat	Common bunt or stinking smut	carboxin Vitavax 34	Vitavax instruction on label. Must	Use according to be applied as a slurry treatment for complete coverage.
		carboxin + captan	Vitavax 20-20	
		carboxin + thiram	Vitavax 200	
		PCNB	Gustafson LT-2N	
		tebuconazole +	Raxil-thiram thiram	
		difenconazole + mefanoxam	Dividend XL RTA	Apply 5-10 fl oz/100 Ib of seed.
		difenconazole + mefanoxam	Dividend Apron XL	Apply 0.5-1.0 fl oz/ 100 Ib of seed.
Wheat	Common root rot complex	imazalil	Gustafson Flo-Pro IMZ	Use according to label instructions.
		difenconazole + mefanoxam	Dividend XL RTA	Apply 5-10 fl oz/100 1b of seed.
		difenconazole + mefanoxam	Dividend Apron XL	Apply 0.5-1.0 fl oz/ 100 Ib of seed.

Contd...

Wheat	Powdery	tridimenol	Baytan 30 Flowable	This fungicide will prevent the mildew overwintering of powdery mildew and leaf rust in susceptible cultivars. Rate is 1.5 fl oz/100 Ib of seed. It is only registered for available for farmer application. Caution: Treated seed should not be planted late in planting season or planted more than 1.5 inches deep. Use according to label instructions
Wheat	Leaf rust	difenconazole + mefanoxam	Dividend XL RTA	Apply 5-10 fl oz/100 Ib of seed.
		difenconazole + mefanoxam	Dividend Apron XL	Apply 0.5-1.0 fl oz/ 100 Ib of seed.
Wheat barley	Aphids (Barley Yellow Dwarf)	imidazolidinimine	Gaucho 480 Flowable	Apply 1.0 fl oz/100 Ib of seed for fall aphid control.
	Hessian fly			Apply 1.5-2.0 fl oz/100 Ib of seed for Hessian fly control.

Contd...

Leaf Rust	Apply fungicide if rust covers 1 to 3% of area of upper leaves
Powdery Mildew	Apply fungicide if powdery mildew covers 5 to 10% of area of upper leaves
Septoria	Apply fungicide if 25% of the indicator leaves (described below have one or more lesions.

Growth Stage	**Indicator Leaves**
Jointing-until flag leaf emergence:	4th and 5th leaf below the flag
Flag leaf emergence until late boot:	
Late boot until start of flowering:	3rd leaf below the flag
Flowering until early milk stage:	2nd leaf below the flag 1st leaf below the flag

Crop	Disease	Fungicide Common Name	Fungicide Trade Name	Formulated Rate/acre	Remarks
Wheat and Barley	Powdery mildew on wheat and barley	Propico-nazole	Tilt 3.6 EC	4.0 fl oz	Apply only when powdery mildew covers 5-10% of area of fully expanded upper leaves. Do not apply more than 4 fl oz/a per year or after Feekes' GS 8 (flag leaf fully emerged). Do not graze or feed livestock treated forage or cut he green crop for hay or silage. After harvest, straw may be used for bedding. Follow all label instructions.

Contd...

Wheat and Barley	Leaf rust on wheat and barley	Propic-o nazole	Tilt 3.6 EC	4.0 fl oz	Apply only when rust pustules cover 1-3% of area of fully expanded upper leaves. Do not apply more than 4 fl oz/a per year or after Feeke's GS 8 (flag leaf fully emerged). Do not graze or feed livestock treated forage or cut the green crop for hay or silage, after harvest straw may used for bedding. Follow all label instructions.
Wheat	Leaf Rust	azoxyst-robin	Quardris oz	6.2-10.8 fl	Integrate into an overall disease management strotegy that includes proper selection of varieties with disease tolerance, proper timing and placement of irrigation, removal of plant debris in which inoclum overwinters, plant residue management and crop rotation. Resistance management: Do not make more than two application of Quadris per acre per. Application directions Quadris should be applied prior to or in the early stages of disease development. Application may be made at any time, immediately after jointing (Feekes 6) up to late head emergence (Feekes 10.5) A crop oil concentrate adjuvant may be added at 1.0% V/V to optimise efficacy. Do not hervest treated wheat for Forage apply later than Feekes growth stage 10.5. Do not apply more than 0.77 quarts product/acre/season (0.4 lb ai/a). Do not apply within 14 days of harvest for hay. Do not apply within 45 days of harvest for rain and straw.

Contd...

Barley and rye	Barely scald, Net, blotch	Propic-onazole	Tilt 3.6 EC	4.0 fl oz	Do not graze or feed livestock treated forage or cut the green crop for hay or silage. After harvest, straw may be used for bedding. Follow all label instructions.
Wheat	Leaf and glume blotch and Tan spot	Propico-zole	Tilt 3.6 EC	4.0 fl oz	Tilt: Do not graze or feed livestock treated forage or cut the green crop for hay or silage. After harvest, straw may be used for bedding. Follow all label instruction.
Wheat	Leaf and glume blotch and tan sopt	azoxyst robin	Quardris fl oz	6.2-10.8	Quadris should be integrated into an ovreall disease management strategy that includes proper selection of varieties with disease tolerance, proper timing and placement of irrigation, removal of plant debris in which inoculum overwinters, plant residue management and crop rotation. ***Resistance Management:*** Do not make more than two application of Quadris should be applied prior to or in the early stages of disease development. Application may be made at any time, immediately after jointing (Feekes 6) up to late head emergence (Feekes 10.5) A crop oil concentrate adjuvant may be added at 1% V/V to optimise efficacy. Do not apply later than Feeks Growth Stage 10.5 Do not harvest treated wheat for forage. Do not apply more than 0.77 quarts product/acre/season (0.4 lb ai/a) Do not apply within 14 days of harvest for hay. Do not apply within 45 days of harvest for grain and straw.

Importance of Plant Afflictions

Plant disease are important because of the loss they cause to the grower. The loss can occur in the field or in the store and at any time between sowing and consumption of the harvest standing crop in the field is attacked by a disease and plants start drying or their capacity to yield satisfactorily is reduced and sometimes creating conditions of food shortage and famine.

The late blight of potato, a disease caused by the fungus *Phytophthora infestance) is* a famous example of what a plant disease can do to change the course of history in 1845. This disease destroyed the potato crop of Ireland where potato constituted the staple diet of the majority in rural areas with Ireland, England and certain parts of continental Europe. Where potato was the staple food of population. There was famine in these countries especially Ireland and about 2 million people died of hunger and many more became diseased due to physical weakness. Many migrated to other land like North-America.

Wheat rust has been another disease that has appeared in epidemic form from time to time in many countries. This disease has forced the farmers in many part of the world to change their cropping pattern. Wheat has been replaced by corn (Maize) or rye because it was regularly destroyed by rust. In the year of the second world war (1943) Bengal had to face a serious famine due to attack of *Helminthosporium* leaf spot *(Helminthosporium oryzae).* Likewise, in the middle of nineteenth century coffee rust. *(Hemileia zeastatrix)* attack the coffee plantation in Ceylon (Sri Lanka) which was the biggest producer and exporter of coffee. During 1867-71 the coffee production was reduced to almost half and by *1893* the export was reduced to *93%*. The planters were forced to cut down coffee plants and shift to tea plantation.

These are only few examples, however, ordinarily, plant disease affect every crop, every year and in all parts of the world. This result in huge loss of potential production and money. Accurate data regarding the losses are not available from all countries because of difficulties in assessment but on the basis of number of plants affected and extraordinary decrease in yield.

In India the losses due to rusts of wheat can range from 8-10%, loose smut of wheat is estimated to cause 2-3% losses. Few years back ear cockle of wheat, a nematode disease, was reported to cause a loss of 50% in U.P. Different smut of Jowar, red rot, potato mosaic, sandal spike disease, bunchy top of banana, rice blast, blight, etc. have caused huge losses in India.

Reasons of Plant Diseases: A pathogen is always, associated with a disease. The word "Pathogen" can be broadly defined as any agent or factor that incites "Pathos" or disease in an organism. Thus in strict sense the pathogens do not necessarily belong to living or animal groups. They may be non-living or in between the living and non-living. The pathogens are thus grouped under the following categories:

(a) *Abiotic Factors:* These include mainly the deficiencies or excess of nutrients, light, moisture, aeration, abnormalities in soil conditions, atmospheric impurities, etc. Examples of diseases caused by abiotic factors are: mango tip rot or fruit necrosis, "Khaira" disease of rice, hollow and black heart of potato.

(b) *Mesobiotic Causes:* These are disease incitants which are neither living nor non-living. They are considered to be on threshold of life.

   (1) *Viroids:* Naked infectious strands of nucleic acid-spindal tuber of potato, citrus exocortis.

   (2) *Viruses:* These are infectious agents made up of single type of nucleic acid (RNAor DNA) enclosed in a protein coat;

Instances: leaf curl of tomato and chilli, etc.

(c) *Biotic Causes:* This category includes diseases caused by animate or living or cellular organism.

Prokaryotes

(a) *Mollicutes:* These are wall less prokaryotes that include mycoplasma-like organism (MLOs) and spiroplasmas.

Instances: Grassy shoot of sugarcane, little leaf of brinjal, sandal spike, papaya bunchy top.

(b) *Rickettsia like Bacteria (RLBs):* These are very small, sometimes sub-microscopic, walled bacteria causing such diseases as citrus greening.

(c) *True Bacteria:* Example of disease will of potato, soft rot of potato, citrus canker, etc.

Eukaryotes

(a) *Fungi:* Potato wart, cabbage club root, potato late blight, downy mildews and powdery mildews, rusts and smuts, red rot of sugarcane, etc.

Nearly 70% of disease in any plant are incited by fungi.

(b) *Protozoa:* Hartrot of coconut palm, phloem necrosis of coffee.

(c) *Algae:* Red rust of mango, papaya.

(d) *Nematodes:* Root-knot of vegetables crops, molya disease of wheat and barley, ear cockle of wheat, citrus decline, etc.

Classification of Plant Afflictions

These diseases may be infectious or non-infectious.

(1) *Non-infectious or Non-parasitic Diseases:* These are diseases (disorder) with which no animate, virus or viroid pathogen is associated. Therefore, they remain non-infectious and can not be transmitted from a diseased plant to another plant. Since no parasite is associated with these disease they are also known as non-parasitic diseases.

(2) *Infectious Diseases:* These diseases are always infectious, sometimes contagious and are transmitted from diseased to healthy plants. These diseases are caused by parasitic pathogens or viruses.

Infectious diseases are often classified according to their occurrence in following groups:

(a) *Endemic Disease:* The word "endemic" means prevalent in and confined to, a particular country, district or location and is applied to diseases of plants and animals including man. These diseases are natural to one country or part of the earth, when a disease is more or less constantly present from year to year in a moderate to severe form, in a particular country or part of earth it is called endemic to that area.

(b) *Epidemic or Epiphytotic Disease:* The term "epidemic" is derived from a Greek word meaning "among the people" and in the true sense applies those diseases of plants humans/animals which appear very virulently among a large section of the population.

To carry the same sense in the case of plant disease the term "epiphytotic" was coined. An epiphytotic disease is one which occurs widely but periodically. It may be present constantly in, the locality but assumes serious proportions only on occasions.

(c) *Sporadic Disease:* Sporadic diseases are those diseases which occur at very irregular intervals and locations and in relatively few instances.

- When a disease is prevalent throughout the country, continent or the world it is known as endemic disease.
- For practical purpose the diseases are also classified according to their symptoms.
- According to symptoms the disease may be rusts, wilts, smuts, blights, cankers, mildews, root rot, fruit rots, leaf spots, etc.
- According to host plants the diseases may be grouped as cereals disease, forage crop disease, flax diseases, root crop diseases, plantation crop diseases, etc.

- When a pathogen survives and spreads through soil, it is known as soil-borne disease.
- However when a disease spreads through seed. It is known as seed-borne disease.
- When the disease dispersal through the agency of air the pathogen is known as air-borne disease.

As is discussed in the introductory para, the primary aim of the plant pathologist is to control the disease and increase the productivity of the crop.

In order to devise suitable control measures one has to throughly study the cause of plant disease its etiology and factors favourable for its dispersal, infection and Epidemiology.

Once the life cycle of the pathogen is completely understood, the pathologist would like to hit at the weakest link of its life cycle through immunisation or prophylatic means to achieve its control.

Plant Disease Control: Its Principles: Control practices are desirable only when the cost in terms of money and efforts is materially less than the loss expected from the disease so that the grower is ensured a margin of profit from plant protection measures since economically and ecologically it is not possible to completely eliminate a pathogen from the ecosystem, therefore, the word "control" is substituted with the word management. The disease management or control involves the following steps.

Immunisation

(Possession of qualities by the host that do not permit establishment of infection).

1. *Resistance:* Development of resistant varieties.
2. *Chemotherapy:* Development of biochemical resistance through the application of antibiotics and systemic fungicides and also through modification of host nutritions.

Prophylaxis (Preventive Measures)

(1) *Protection:*

(a) *Mechanical:* By removing the diseased material or seed from the healthy lot before sowing.

(b) *Physical:* By heat treatment, etc.

(c) *Cultural:* By modifying cultural operations like time of sowing depth, crop rotation, modification of environment, sanitation, roguing of diseased plants, production of disease free seed or propagated material.

(2) *Eradication:*

(a) *Crop Rotation:* Against those pathogens which persist for longer in the soil.

(b) *Sanitation:* To remove the diseased plants or residue from the field.

(c) *Alternate Host Eradication:* To eliminate the primary inoculum produced on them.

(d) *Plant Therapy:* Through fungicides heat and also surgery (in case of tree).

(e) *Chemicals:* Through fungicides and antibiotics.

(f) *Biological Control:* Through the activity of other micro organisms.

(3) *Legislation:* Aims at preventing entry of pathogen from infested area to non-infested areas.

Such regulations are called quarantine. The governments may have international or domestic quarantine against a particular disease.

Integration of Plant Disease Management Practices (IPDMP): The four basic requirement for management of plant diseases are: clean and healthy seeds, clean field or pathogen free soil, prevention of entry of and infection by a pathogen in a standing crop and precautions during harvesting and storage of the produce.

The crop rotation is another precaution during growth of the crop periodical roguing of disease plant.

Plant Pathology Facts

1.	Father of Pathology	Anton de Berry
2.	Father of Mycology	P. A. Micheli
3.	Father of Nematology	N. A. Cobb
4.	Father of Indian Pathology	E. J. Butler
5.	Bacterial technique was given by	Dr. Robert Koch
6.	Mycoplasma disease was first reported by	Do. *et.* al. and Ishiil *et. al.*
7.	First plant parasitic bacteria was reported by	T. J. Burill
8.	First plant parasitic bacteria was	Fire blight of apple
9.	First plant parasitic nematode was reported by	Needham
10.	First plant parasitic nematode was	Anguina *tritici*
11.	"Stanley" did crystallisation of	Viruses
12.	Late blight of potato caused by	*Phytophthora infestance*
13.	Brown spot of rice caused by	*Helminthosporium oryzae* caused Bengal femine (1845)
14.	Coffee rust occurred in	Sri Lanka 1867
15.	Maximum number of plant parasitic	Tylenchida order nematode belongs
16.	Loose smut of wheat is	Internally seed borne *(JRF 2000)*
17.	Indian phytopathological society was established by	E. J. Butler
18.	Antibiotic penicillin was discovered by	Alexander Flemming

19. Example of obligate parasite is	*Albugo candida*
20. Green ear disease of bajra is a	Soil and air borne diseases
21. Life cycle of wheat rust is given by	K. C. Mehta
22. Systemic fungicide is discovered by	Ven Schleming and Kulka
23. Yellow vein mosaic virus was reported by	Uppal *et. al.*
24. Tomato mosaic virus was reported by	Adolph Mayer
25. Taphrina causes	Peach leaf curl
26. Endoparasitic nematodes is	Meloidogyne Lichen (Algae + Fungi).
27. Example of symbiosis is	Mycorrhiza (Fungi + Tree roots)
28. Citrus canker	*Xanthomonas campestris var. citri*
29. Molya disease of wheat due to	Nematode
30. Black tip of mango is caused by	Boron deficiency
31. Botulism is caused by species of	*Clostridium*
32. Agar-agar is produced from	*Gelidium*

BIBLIOGRAPHY

Abbasi, S.A. : *Wetlands of India : Ecology and Threats,* Discovery Publishing House, New Delhi, 1997.

Adkins, R.M. : *Annual Review of Ecology and Systematics,* A. R., London, 1993.

Akhtar, R. : *Disease Ecology and Health,* Rawat Publications, Jaipur, 2001.

Alison, J. : *Spatial Ecological : Economic Analysis for Wetland Management,* Cambridge, London, 2004.

Anthony, J. : *Handbook of Ecological Restoration,* Cambridge, London, 2002.

Antonovics, Janis : *Integrating Ecology and Evolution in a Spatial Context,* Cambridge, London, 2001.

Banerjee, B.N. : *Environmental Pollution and Bhopal Killings,* Gyan Books, New Delhi, 1987.

Barker, G. : *Physiological Plant Ecology,* Cambridge, London, 2002.

Berger, W.H. : *Abrupt Climatic Change,* D. Radiel Dordrecht, Boston, 1990.

Bhargava, Gopal : *Ecological Politics : Different Dimensions,* Gyan Books, New Delhi, 2002.

——— : *Marine Ecosystems,* Gyan Books, New Delhi, 2001.

Bhusan, Bibhuti : *Social Ecology of Forest Resources,* Gyan Books, New Delhi, 2004.

Billihgs, W.D. : *Plant, Man and the Ecosystem,* MacMillan, London, 1971.

Bookchin, M. : *The Philosophy of Social Ecology,* Rawat Publications, Jaipur, 1996.

Chandra, Ramesh and Aneja, Ritu : *Corporate Governance for Sustainable Environment,* Gyan Books, New Delhi, 2004.

Colin, J. : *Issues in Environmental Economics,* Blackwell, London, 2002.

Crawford, T.J. and Hewitt, G.M. : *Genes in Ecology,* Cambridge, London, 1992.

Crosby, A.W. : *Ecological Imperialism : The Biological Expansion of Europe,* Cambridge, London, 2000.

Dakshinamurthi, C. : *Water Resources of India and their Utilization in Agriculture,* Oxford University Press, New Delhi, 1973.

Dass, Sujata K. : *Environment Challenges and Sustainable Future,* Gyan Books, New Delhi, 2003.

Dasseman, R.F. : *Environmental Conservation,* John Wiley, New York, 1976.

Dave, M. : *Urban Ecology and Levels of Development,* Rawat Publications, Jaipur,

David, Khemchand : *Advanced Dictionary of Ecology and Environment,* Dominant Publishers, New Delhi, 2000.

DeVere, Burton : *Ecology of Fish and Wildlife,* Delmar, London, 1996.

Dickinson, L. : *Ecology and Evolution of Cooperative Breeding in Birds,* Cambridge, London, 2004.

Dodson, S. I. : *Ecology,* Oxford, New York, 1998.

Dutta, Ashish : *Biodiversity and Ecosystem Conservation,* Gyan Books, New Delhi, 2001.

Dwivedi, S.D. : *Naturopathy : For Perfect Health,* Gyan Books, New Delhi, 2002.

Eric, R. Pianka : *Evolutionary Ecology,* Harper Collins, London, 1994.

Fritz, L. Knopf : *Ecology and Conservation of Great Plains Vertebrates,* Springer, London, 1997.

Gaan, Narottam : *Environmental Security Concept and Dimensons,* Gyan Books, New Delhi, 2004.

Gary, A. Klee : *The Coastal Environment : Toward Integrated Coastal and Marine Sanctuary Management,* Prentice Hall, London, 1999.

George, A. Maul : *Climatic Change in the Intra : Americas Sea,* Arnold, New Jersey, 1993.

Gerry Closs, Barbare Downes and Andrew, Boulton : *Freshwater Ecology : A Scientific Introduction,* Biackwell, London, 2004.

Gordon, D. : *World Population Problems,* Park Press, Campaign, 1984.

Grafton, Q. : *Economics of the Environment,* Blackwell, London, 2004.

Gupta, K. : *Energy and Environment in India : A Study of Energy Management,* Gyan Books, New Delhi, 2002.

Hourani, G.F. : *Arab Seafaring in the Indian Ocean and Early Medieval Times,* Princeton University Press, Princeton, 1951.

Husain, M. : *Evolution of Geographical Thought,* Rawat, Jaipur, 1995.

Hussen, A. : *Principles of Environmental Economics,* Routledge, New York, 2004.

Hutchings, J. : *The Ecological Consequences of Environmental Heterogeneity,* Cambridge, London, 2000.

Ian, D. Whyte : *Climating Change and Human Society,* Arnold, New Jersey, 1995.

Ian, F. Spellerberg : *Monitoring Ecological Change,* Cambridge, London, 2005.

James, K. Lein : *Integrated Environmental Planning,* Blackwell, London, 2003.

James, M. Bullock : *Dispersal Ecology,* Cambridge, London, 2002.

John, E.FA : *Evolution and Ecology of Macque Societies,* Cambridge, London, 1996.

John, N. Rayner : *Dynamic Climatology : Basis in Mathematics and Physics,* Blackwell, London, 2001.

Joshi, M.V. : *Theories and Approaches of Environmental Economics,* Atlantic Publishers, New Delhi, 2004.

Kaur, Karanjot : *Green Revolution : Ecological Implications,* Dominant Publishers, New Delhi, 1999.

Khanna, D.R. : *Microbial Ecology : A Study of River Ganga,* Discovery Publishing House, New Delhi, 2004.

Kumar, Chattopadhyay : *The Desperate Delta : Social Ecology of Sunderbans,* Gyan Books, New Delhi, 1995.

Kumar, Shashi : *Human Ecology for Globalization : Human Ecology in Action,* Atlantic Publishers, New Delhi, 2003.

——— : *Human Ecology for Globalization : New Dimensions for Economic Growth,* Atlantic Publishers, New Delhi, 2004.

Madan, Mohan : *Ecology and Development,* Rawat Publications, Jaipur, 2000.

Martin, T.E. : *Ecology and Management of Neotropical Migratory Birds,* Oxford, New York, 1995.

Mash, G. P. : *Man and Nature Physical Geography as Modified by Human Action,* Charles Scriber, New York, 1964.

Max, N. : *The Environmental Revolution,* McGraw-Hill, New York, 1970.

Michael, Moss : *Issues and Perspectives in Landscape Ecology,* Cambridge, London, 2005.

Moore, Francis : *Environment and Society,* Dominant Publishers, New Delhi, 2003.

Mukherjee, Roma : *Environmental Management and Awareness Issues,* Sterling Publishers, New Delhi, 2002.

Muthiah, S. : *An Atlas of India,* Oxford University Press, New York, 1992.

Myers, H. : *Ecology and Control of Introduced Plants,* Cambridge, London, 2003.

Nadarajah, M. : *Culture, Gender and Ecology : Beyond Workerism,* Rawat Publications, Jaipur, 1999.

Nigel, J.R. : *Mountains at Risk : Current Issues in Environmental Studies,* Manohar Publishers, New Delhi, 1995.

Nigel, Webb : *Ecological Reviews,* Cambridge, London, 2005.

Overbeck, J. : *Microbial Ecology of Lake Plubsee: Ecological Studies,* Springer, London, 1994.

Pandey, Mahendra : *Environment Pollutants & Women's Health,* Dominant Publishers, New Delhi, 2003.

Paul, A. : *Introduction to Ecology,* John Wiley and Sons, New York, 1973.

Peter, J. Grubb and John, B. Whittaker : *Toward a More Exact Ecology,* Cambridge, London, 1989.

Puri, G. : *Essential Soil Science : A Clear and Concise Introduction to Soil Science,* Blackwell, London, 2002.

Rampinot, M.R. : *Climate: History, Periodicity and Predictability,* Von Nostrand Reinholt, New York, 1987.

Ravi, V.S. : *The Living Adventure : Nature and Biology,* Manohar Publishers, New Delhi, 2001.

Raza, M. : *Development and Ecology,* Rawat Publications, Jaipur, 2000.

Ring, T. Carde and Jocelyn, G. Millar : *Advances in Insect Chemical Ecology,* Cambridge, London, 2004.

Rob, H.G. Jongman : *Ecological Networks and Greenways,* Cambridge, London, 2004.

Sarah, J. Woodin and Mick, Marquiss : *Ecology of Arctic Environments,* Cambridge, London, 1997.

Sheth, P. : *Environmentalism : Politics, Ecology and Development,* Rawat Publications, Jaipur, 1997.

Shukla, S.K. and Srivastava, P.R. : *Ecology and The Environment,* Commonwealth Publishers, New Delhi, 1992.

Sinha, R.K. : *Eco-Politics and Development in India,* Commonwealth Publishers, New Delhi, 1996.

Stamou, G.P. : *Arthropods of Mediaterranean : Type Ecosystems,* Springer, London, 1998.

Thomas, R. : *Man's Impact on the Environment*, McGraw-Hill, New York, 1988.

Thomson, T. : *Introductory Essay to Flora Indica*, MacMillan, London, 1855.

Trivedi, P.R. and Dudarshan, K.N. : *India Ecology and Environment*, Commonwealth Publishers, New Delhi, 1994.

Usher, Michael : *Ecology, Biodiversity, and Conservation*, Cambridge, London, 2005.

Vayda, A.P. G. : *Man in the Pacific Islands: Essays on Geographical Change in the Pacific*, Oxford University Press, New York, 1972.

Venus, W. : *Kanpur City: A Study in Environmental Pollution*, Tara Book Agency, Varanasi, 1982.

Vogt, K.A. : *Ecosystems : Balancing Science with Management*, Springer, London, 1997.

Walter, G.H. : *Insect Pest Management and Ecological Research*, Cambridge, London, 2003.

Wilfried, B. : *Atmospheric Pollution*, McGraw Hill, New York, 1972.

William, W. : *Integrating Landscape Ecology into Natural Resource Management*, Cambridge, London, 2002.

Yadav, P.R. : *Environmental Ecology*, Discovery Publishing House, New Delhi, 2003.

Yeates, M. : *Plant Ecology*, Harper & Row, New York, 1980.

Zelinsky, W. : *A Study of Ecology*, Oxford University Press, New York, 1970.

Zuckerman, Ben : *Human Population and the Environmental Crisis*, Jones & Bart., New Jersey, 1996.

INDEX

C

D

E

F

G

H

I

J

K

L

M

N

S

T

U

V

W

X

Y

Z

❑❑❑